Digital Video Broadcasting: Technology, Standards, and Regulations

For a complete listing of the *Artech House Digital Audio and Video Library,* turn to the back of this book.

Digital Video Broadcasting: Technology, Standards, and Regulations

Ronald de Bruin
KPMG

Jan Smits
Eindhoven Centre of Innovation Studies (ECIS)
Eindhoven University of Technology
The Netherlands

Artech House
Boston • London

Library of Congress Cataloging-in-Publication Data
Bruin, Ronald de.
    Digital video broadcasting : technology, standards, and regulations /
Ronald de Bruin, Jan Smits.
    p.   cm.
    Includes bibliographical references and index.
    ISBN 0-89006-743-0 (alk. paper)
    1. Digital television.   2. Television broadcasting.   I. Smits, Jan
II. Title.   IV. Series
TK6678.B78   1998
384.55—dc21   98-51785
                                                CIP

British Library Cataloguing in Publication Data
Bruin, Ronald de
    Digital video broadcasting : technology, standards, and regulations
    1. Digital television
    I. Title   II. Smits, J. M., 1953–
    621.3'88

    ISBN   0-89006-743-0

Cover design by Lynda Fishbourne

© 1999 ARTECH HOUSE, INC.
685 Canton Street
Norwood, MA 02062

International Standard Book Number: 0-89006-743-0
Library of Congress Catalog Card Number: 98-51785

10 9 8 7 6 5 4 3 2 1

To Susanne whose eyes mirror the true values of life.
Ronald de Bruin

To Saskia who always encourages me to explore new frontiers in science and law and to Jan-Paul who has brought so much joy and the conviction that the future will be bright.
Jan Smits

Contents

Contents

Foreword

Many books and articles on the transition of television from analog to digital transmission have been published in recent years. This transition is still in its infancy, but it will inevitably take place in the years to come. Due to digitization, the existing barriers between audio, video, and data generation and transmission will cease to exist, and the face of what we today call television will radically change.

This book is for individuals who want to obtain inside knowledge on *digital television* (DTV). Without pretending to be exhaustive, it provides an overview of DTV technology, standards, and regulation with an emphasis on the development of the standards generated by the European Project for *digital video broadcasting* (DVB). In addition, this book compares the various DVB standards for cable, satellite, and terrestrial transmission and describes European, American, and Japanese regulations. DVB started with broadcasting and, as discussed in this book, gradually moved to the specification of return channels in the telecommunications domain and finally into specifications for interactive and data broadcasting.

The DVB Project is recognized throughout the world as an unprecedented success in standardization and rapid implementation in the market. The active involvement of all players in the television value chain has been essential to this triumph. These television industry entities, which work in the business environment rather than in standardization bodies only, wisely decided that commercial requirements had to precede technical specifications. Early involvement of regulators and standardization bodies has proven to be essential to success within the geographical area of Europe and has created the basis for DVB's ability to spread out the DVB specifications into de facto standards in many parts of the world outside Europe.

Theo H. Peek,
Chairman of the DVB Steering Board and General Assembly
Eindhoven, The Netherlands
25 June 1998

Preface

Technological developments should not be regarded as exogenous determining factors but rather as the product of activities and relationships within society as a whole. Beside the technical factors involved, scientific, economic, market, political, and legal factors can determine the establishment of technologies in society. This book aims to provide an overview of these aspects with regard to DTV and to help explain how DTV, including conditional access, can be successfully embedded in society.

We believe that the European initiatives on DVB will play an important role in the establishment of DTV throughout the entire world. Thus, this book focuses mainly on the European (technological) developments in DVB. To place these developments in a global context and to complete the required overview, the U.S. and Japanese policies and regulations are discussed as well. This overview enables an analysis of how DTV services can successfully be embedded in society. Due to the many emerging aspects of the development of DVB, much emphasis is given to the provision of services via conditional access systems (e.g., pay television). Finally, some (possible) future DVB developments on a mid-term time scale are discussed.

This book is geared towards decision makers, policy makers, managers, and engineers in government, business, and academic institutions involved in media, telecommunications, and the terminal equipment manufacturing industry. It provides a general overview of , as well as specific (technical) information on the different aspects of DTV. Readers will develop an understanding of the establishment of European, Japanese, and U.S. DTV policies, regulations, and market developments, and, from a technology perspective, will be able to compare the several European DVB standards. More academically oriented readers will enjoy the book's technology assessment of the European DVB framework.

Acknowledgments

We are indebted to many people for their information, feedback, and assistance during the development of this book. First of all, we would like to thank Theo Peek, chairman of the DVB Steering Board and General Assembly, for sharing his experience and broad overview. We would also like to thank Eamon Lalor and his staff with the European Commission DGXIII for the stimulating discussions we had in Brussels. Furthermore, we are much obliged to Masaaki Kobashi of MITI and Arjen Blokland of the Dutch Embassy's Scientific Office in Tokyo for providing us with information on Japanese DVB developments. We also thank Theo van Eupen of the Nederlands Televisie Platform for providing us with valuable information on *high-definition television* (HDTV). Moreover, we would like to thank Jan Bons, Telematic Systems & Services B.V., for his contributions and feedback on interactive television systems.

Several individuals helped us to improve our manuscript by providing suggestions and feedback: We would like to acknowledge Peter Anker, Lucas van der Hoek, and Paul van der Pal, all employees of the Dutch Telecommunications and Post Department, and Arch Luther, Artech House's *Digital, Audio, and Video Series* reviewer. Our special gratitude goes to Tessa Halm for her most valuable review of grammar and style.

We have been most happy to receive the assistance of Menno Prins who compiled a professional file, Paul Dortmans who implemented the proper software conversions of our artwork, and Ursula Kirchholtes who constructed the glossary's first outline. Moreover, we would like to thank Paula Verheij for helping to shape some of our texts into the required formats.

Finally, we are grateful to our families and friends for their patience, understanding, and moral support.

Den Haag, Utrecht, June 1998

CHAPTER

1

Contents

History of digital television

1.1 Introduction

In 1883, the French novelist Albert Robida wrote his book *Le Vingtième Siècle* (*The Twentieth Century*) [1], which describes a very particular vision. In the novel, a spectator sits in a comfortable chair in his living room watching life-size pictures of a scene that takes place at another location. These pictures are being projected by what Robida calls a *telephonoscope*. This corresponds closely with the television system as we know it today.

The basic purpose of television systems is to extend the senses of vision and hearing beyond their natural limits. In technical terms, television is the conversion of a scene in motion with its accompanying sounds into an electrical signal, transmission of the signal, and its reconversion into visible and audible images by a receiver [2]. The first television systems were mechanical; later, they became electronic. The next innovation was color television, followed by a high-quality system called *high-definition television* (HDTV). The latest innovation is based on the application of digital techniques, through which the traditional boundaries between media and telecommunications have disappeared. This paves the way for all different kinds of interactive multimedia services.

1.2   Mechanical television

In 1884, the 24-year-old German student Paul Gottlieb Nipkow obtained a patent on the very first television system [3]. This system operates as follows (see Figure 1.1): First, an image is illuminated by a lamp via a lens and a *Nipkow disk*, which has square apertures arranged in a spiral. The rotation of the disk provides a simple and effective method of image scanning. As the disk rotates, the outermost aperture traces out a line from the left to the right across the top of the image. The next outermost aperture traces out another line, directly below and in parallel to the preceding line. After one rotation, the successive apertures have traced out parallel lines, left-to-right, top-to-bottom, so that the whole image has been scanned. The more apertures there are, the more lines there are traced, and hence the higher the level of detail.

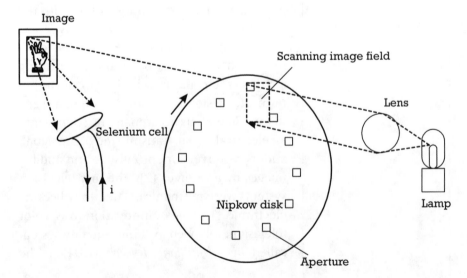

Figure 1.1   The Nipkow mechanical television system.

The reflected light from the image is collected by a selenium cell. (In 1873, it was discovered that the electrical conduction of selenium varied with the amount of illumination. When the intensity of the reflected light varies with the parts of the image, the current in the cell also varies. Hence, the lighter parts of the image are represented by a stronger current than the darker parts.) Finally, at the receiving end a lamp emits more or less light in correspondence with this current. If the same type of disk is used in a synchronized way, the original image can be reproduced. Moreover, it is essential that the disk's rotation is at sufficient speed for the eye to perceive the image as a whole, rather than a sequence of moving points.

In 1895, Perrin and Thomson discovered the existence of the electron. Two years later, a German named K. F. Braun invented a screen that produced visible light when struck by electrons. He designed a cathode-ray tube by which means a beam of electrons could be aimed at the fluorescent screen. In 1904, Englishman J. A. Flemming invented the two-way electrode valve, and in 1906, Lee de Forest added the grid, which enables amplification. It was the Russian scientist Boris Rosing who first suggested using the cathode-ray tube in the receiver of a television system in 1907. At the camera end he used a mirror-drum scanner.

In 1908, Scottish electrical engineer A. A. Cambell Swinton proposed the use of magnetically deflected cathode-ray tubes at both the receiving and camera end. The camera contained a mosaic of photoelectric elements. The back of the camera screen was discharged by a cathode-ray beam. According to Nipkow's principle, the beam scanned the image line by line. This proposal in essence formed the basis of modern television. Nipkow's ideas were too advanced to put into practice at that time. However, he explained his ideas in several publications and in an address to the Röntgen Society of London in 1911.

In 1924, J. L. Baird in Britain used triode amplifiers and replaced the selenium cell by a gas filled potassium photocell. This improved the photocell's response time to changes in the light. In addition, Baird adopted the principle of modulated light from the American D. M. Moore. By varying the electrical input of a neon gas-discharge lamp at the receiving end, it is possible to vary the light intensity of this lamp. Baird used a Nipkow disk for 30 lines and a speed of five images per second, which he later improved to 10 images per second. In 1926, Baird demonstrated the first true television system. Meanwhile, the American C. F. Jenkins experimented with mechanical methods using the Nipkow principle as well. He also replaced the selenium cell but used an alkali metal photo cell instead.

The first television standard was established in 1929. It read, "A screen consists of 30 lines and 1,200 elements" [4]. In 1931, a new standard was defined (48 lines and 25 images per second). By this standard, the limits of the still mechanical display at the receiving end were reached. Table 1.1 chronologically details the evolution of mechanical television systems.

1.3   Electronic television

Swinton already determined that for a good display, quality images need to be analyzed into at least 100,000 and preferably 200,000 elements. The number of elements is approximately equal to the square of the number of lines. This implicates that the mechanical systems using 30 or 48 lines do not meet this requirement by far.

The Russian emigrant Vladimir Kosma Zworykin made a very important step forward in 1923 when he replaced the Nipkow disk with an electronic component. It then became possible to split up the image into

Table 1.1
Developments in Mechanical Television Systems

Date	Development
1873	Electrical conduction of selenium
1884	Nipkow disk
1895	Discovery of the electron by Perrin and Thomson
1897	Cathode-ray tube by K. F. Braun
1904	Two-way electrode valve by J. A. Flemming
1906	Grid (amplification) by Lee de Forest
1908	Magnetically deflected cathode-ray tube by A. A. Cambell Swinton
1913	Potassium photocell by German research
1917	Modulated light by D. M. Moore
1924	Baird system
1925	Jenkins system
1929	First television standard (30 lines, 1,200 elements)
1931	Television standard (48 lines, 25 images per second)

many more lines, which allowed a higher level of detail without increasing the number of scans per second. Moreover, the tube sensitivity was increased by a unique "storage" feature. The image was stored during the time that elapsed between two electronic scans. In 1925, Zworykin applied for a patent, and in 1933 he put his design into practice. With his *iconoscope,* he proved the theoretical ideas of Swinton.

In Great Britain, the first fully functional electronic television system was demonstrated in 1935 by a television research group from the *Electric Musical Industries* (EMI) under Sir Isaac Shoenberg. The camera tube, known as the *Emitron,* was an advanced version of the iconoscope. At the receiving end, an improved high-vacuum cathode-ray tube was used. Shoenberg proposed a standard for 405 lines with 50-Hz interlaced scanning to allow the scanning of 25 images per second without any flickering. Interlaced scanning implicates that an image is scanned twice (see Figure 1.2). First, scanning field A including the odd lines is scanned and then scanning field B with the even lines is scanned. At the receiving end, both scanning fields are combined (i.e., displayed sequentially). In effect, the picture repetition rate is doubled, which results in a more fluent picture on the screen, while the scanning rate remains the same.

After government authorization, Schoenberg's standard was adopted by the *British Broadcasting Corporation* (BBC). In 1936, this led to the

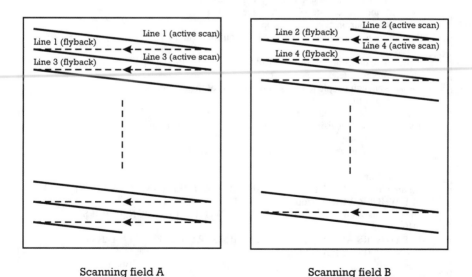

Scanning field A   Scanning field B

Figure 1.2   Principle of interlaced scanning.

launch of the first public television service (high-definition public television service) in the world. In 1937, a standard for 441 lines with 50-Hz interlaced scanning was introduced in Germany. After the United Kingdom, regular television broadcasting began in France in 1936. Later, France began using 819 lines and 50-Hz interlaced scanning. On April 30, 1941, regular television broadcasting began in the United States, where the first mass market for television receivers arose. Since 1927, the Philips company in the Netherlands had been working on the development of television systems as well [5]. Inspired during his visit to the United States in 1948, Bouman from Philips sent a telegram (see Figure 1.3) to Rinia, who was responsible for Philips' television activities. In the Netherlands, regular television broadcasting started on October 1, 1951. Japan followed in February 1953.

C.O.B.-GRAM

URGENT

Date: 5/3/48
Time: 9.15

Lopes Cardozo - Try to stop developmentwork broadcast-
receivers and concentrate all efforts on television stop
Television is our biggest chance stop
Protelgram is allright but we have to work on followup like
hell! Stop
Write on all doors and walls and blackboards TELEVISION stop
Make everybody televisioncrazy stop
We have enough people to do the job but most of them work on the
wrong items stop
There really is only one item: TELEVISION stop
The only actual televisionfront we have at the moment is right
here in U.S.A. stop
We are able to force it if we are ready to fight AND TO KEEP
FIGHTING! Stop
Mobilise Eindhoven please stop
No time to lose TELEVISION IS MARCHING ON HERE AND FROM HERE OVER
THE WHOLE WORLD stop
The only question is: WHO MARCHES ON WITH TELEVISION, PHILIPS OR
THE OTHERS? Stop
THE OTHERS ARE ALREADY MARCHING! P H I L I P S   E I N D H O V E N,
T A K E   T H E   L E A D ! full stop

Bouman

Figure 1.3 Bouman's telegram to Rinia in 1948.

With the introduction of public broadcasting services, the need for standardization concerning the number of lines and frames per second increased. The number of lines is subject to an effective tradeoff between an adequate picture definition and a technically and economically acceptable bandwidth. Another aspect of standardization was the picture repetition rate. The United States (and later Japan) adopted a picture repetition rate of 30 pictures per second, because this rate was easy to derive from its electrical power supply, which is provided at a frequency of 60 Hz. In Europe, the electrical power is provided at 50 Hz. Hence, the picture repetition rate became 25 in Europe. This led to two standards in the world: the U.S. standard for 525 lines per picture at 30 pictures per second used in North America, South America, and Japan and the European standard for 625 lines at 25 pictures per second used in Europe, Australia, Africa, and Eurasia. Table 1.2 details the evolution of electronic television systems.

1.4 Color television

The development of color television did not immediately follow the beginning of regular television broadcasting. In fact, the first ideas for a color television system lead back to a German patent dating from 1904. In 1925, Zworykin filed a patent for an electronic color television system.

Table 1.2
Developments in Electronic Television Systems

Date	Development
1925	Patent electronic television system by V. K. Zworykin
1933	Iconoscope by V. K. Zworykin
1935	British television standard (405 lines with 50-Hz interlaced scanning)
1936	First public television service by the BBC in the United Kingdom
1936	Regular television broadcasting in France (later using 819 lines with 50-Hz interlaced scanning)
1937	German television standard (441 lines with 50-Hz interlaced scanning)
1941	Regular television broadcasting in the United States (525 lines with 60-Hz interlaced scanning)
1951	Regular television broadcasting in the Netherlands (625 lines with 50-Hz interlaced scanning)
1953	Regular television broadcasting in Japan (525 lines with 60-Hz interlaced scanning)

However, it was Baird who demonstrated the first working (mechanical) color television system in 1928. Baird's system used a Nipkow disk with three spirals, one for each primary color (red, green, and blue). While rotating, this system produced a sequence of primary color signals. In 1929, H. E. Ives managed with a mechanical system to transmit the three primary color signals simultaneously via three different channels between Washington, D.C., and New York City. Later that year, his colleague Frank Gray patented a color television system that was based on transmission of the three primary color signals via one and the same channel.

Two basic principles apply to transmission of the three primary color signals via one channel. The primary colors can be transmitted sequentially on a frame-by-frame basis. Alternatively, the primary colors can be transmitted simultaneously, which allows a more efficient use of signal bandwidth. The latter also allows compatibility with black-and-white television systems. In 1938, G. Valensi of France applied for a patent for a color television system that was compatible with black-and-white television systems. Although his system has not been adopted in practice, his ideas on compatibility have proven to be very important.

The first color television service started in the United States in 1951 by means of the abortive frame sequential system. In 1953, the *National Television Systems Committee* (NTSC) in the United States developed a fully compatible system using simultaneous transmission. This NTSC system still forms the basis of color television systems today. As such, this system applies a combined transmission of the image's brightness information and color information. The image's brightness concerns the level of detail and sharpness. This type of information can be interpreted by a black-and-white receiver, which does not use (nor need) the color information. A color television system, however, makes use of both types of information. The first public television broadcasting using NTSC began in the United States in 1954, followed by Japan in 1960.

The NTSC system showed sensitivity for certain distortions caused during transmission and signal processing. These distortions resulted in hue errors, which could be only partially remedied [6]. For this reason, the system's acronym is sometimes said to represent "never the same color." In 1957, Henri de France developed his *système Électronique couleur avec mémoire* (SECAM), with which he tackled the hue error problem.

With the same result, the German W. Bruch modified the NTSC system in 1961 and developed the *phase alternation line* (PAL) system.

In 1967, public television broadcasting using SECAM started in France and the former Soviet Union. In the same year, public television broadcasting using PAL was launched in Germany and the United Kingdom. Today, SECAM is used in France, Greece, Eastern Europe, and Iran, while PAL is used in the rest of Western Europe and many other countries, including Brazil, Argentina, and China. Table 1.3 chronologically details the evolution of color television systems.

1.5 High-definition television

The term HDTV is almost as old as the first mechanical television systems. It has been used to refer to a kind of ideal system or to that which had not

Table 1.3
Developments in Color Television Systems

Date	Development
1904	Patent color television in Germany
1925	Patent electronic color television system by V. K. Zworykin
1928	First demonstration of a mechanical color television system by J. L. Baird
1929	Transmission of color television images via three separate channels by H. E. Ives
1929	Transmission of color television images via one channel by F. Gray
1938	Patent color television system compatible with black-and-white television system by G. Valensi
1951	First color television service in the United States
1953	NTSC standard (525 lines with 60-Hz interlaced scanning)
1954	Public color television broadcasting in the United States using NTSC
1957	SECAM standard (625 lines with 50-Hz interlaced scanning) by H. de France
1960	Public color television broadcasting in Japan using NTSC
1961	PAL standard (625 lines with 50-Hz interlaced scanning) by W. Bruch
1967	Public color television broadcasting in Germany and the United Kingdom using PAL
1967	Public color television broadcasting in France and the former Soviet Union using SECAM

yet been reached. An important element in this discussion is the number of lines used to represent an image. J. L. Baird called his mechanical 30-line system an HDTV system. Nowadays, the use of the term HDTV has stabilized, with today's HDTV systems using about 1,000 lines. However, the frontiers still have not been reached. In Europe, the term *very* HDTV is used to refer to broadband HDTV with studio quality, and in Japan the term *ultra* HDTV stands for a system with 3,000 lines.

In the mid 1960s, the Japanese Dr. Takashi Fuijo from *Nippon Hoso Kyokai* (NHK) started research on a high-quality television system (i.e., comparable with 35-mm film and CD-quality audio), with the objective of achieving a world standard for program production. In Japan, this development was called *High-Vision* instead of television. In the second half of the 1970s, the first broadcasts took place with a 1,125-line system with 60-Hz interlaced scanning. In 1981, NHK demonstrated a HDTV system developed by Sony in the United States. In addition, NHK developed the analog *multiple sub-Nyquist sampling encoding* (MUSE) transmission standard for satellite services in 1984.

Other parts of the world reacted to the Japanese achievements. In 1981, the *European Broadcasting Union* (EBU) started a Working Party V with the objective of studying HDTV, which in Europe was also called *Cine-Vision*. A year later in the United States, the *Advanced Television Systems Committee* (ATSC) was established. The EBU and the ATSC worked in close cooperation, basing their work on NHK's results. In September of 1983, the *Comité Consultatif International des Radiocommunications* (CCIR) *Interim Working Party* (IWP) formed with the goal of proposing a world standard for program production as well as transmission. In 1985, the CCIR IWP delivered a proposal for a production standard based on 1,125 lines with 60-Hz interlaced scanning. An important factor in the EBU's adoption of this proposal was that NHK had developed a standard 1,125 lines/60 Hz to 625 lines/50 Hz converter. The ATSC adopted the standard as well but defined a wide-screen aspect ratio (screen format) of 15:9 instead of the proposed 16:9 aspect ratio.

However, the EBU underestimated the consumer electronics industry lobby. The European consumer electronics industry's (economical) interests were not sufficiently taken into account, as its products and services are mainly based on 50 Hz. Consequently, in 1985 the European Commission asked its member states to disagree with the CCIR IWP proposal. Moreover, the European Commission decided that a decision on

HDTV would be postponed for at least two more years. This also affected the CCIR plenary meeting held in Dubrovnik in May 1986. A CCIR decision on HDTV was postponed until the next plenary to be held in 1990 in Düsseldorf. Hence, Europe and the United States set sail on separate courses.

Preceding the Dubrovnik meeting, on March 12, 1986 the European consumer electronics industry drew up a memorandum of understanding to develop equipment to support HDTV services within Europe. To achieve this objective, the industry initiated a project within the European *Eureka* research program that worked towards a proposal for a European HDTV system based on 50 Hz. Since the number 95 was assigned to this project, it became known as the *Eureka95* project. Its objectives were the following:

▶ The development of a European proposal for an HDTV program production standard to be presented on the CCIR plenary in 1990. One of the requirements was that the standard be based on 50-Hz but also allow possibilities for 60-Hz countries.

▶ The stimulation of the transmission of HDTV programs by means of the *high-definition multiplexed analog component* (HDMAC) satellite transmission standard to achieve reception with conventional MAC-receivers. (Earlier in 1986, the EBU had already specified the MAC/packet family transmission standards, consisting of CMAC, DMAC, and D2-MAC, as an alternative for PAL and SECAM.)

▶ The construction and demonstration of a complete HDTV chain from program production, to transmission, to the reception and storage of HDTV programs.

▶ Fundamental research on key components of HDTV.

The Eureka95 project was planned between 1986 and 1990. A consortium of about 80 participants, led by Philips and Thomson, worked on proposals for the HDTV system's standards, demonstration, feasibility, and the first prototypes. These proposals concerned an HDTV system with 1,250 lines and 50-Hz interlaced scanning. The participants, which were financially supported by the governments, spent a budget of about 200 million ECU.

The Japanese and subsequent European achievements formed a threat to the position of North American broadcast stations [7]. These information providers depend on finances obtained through interregional commercials. The actual broadcasting is processed by regional partners called *affiliates*. Affiliates broadcast their parent stations' programs and commercials, and, in addition, finance themselves with local or regional commercials. This market structure would be destabilized if the Japanese MUSE or the European HDMAC were used to broadcast stations' programs via satellite directly. Moreover, the market structure in the United States had already changed with the introduction of broadcasting via *cable antenna television* (CATV) networks. The CATV network operators play an important role in the provision of television programs at the local and regional levels. Hence, in 1987 the FCC initiated the development of an HDTV standard for (regional) terrestrial broadcasting.

In 1989, the Eureka95 participants decided to extend the project with a two-year program—from July 1, 1990, to July 1, 1992—that increased the total budget to 625 million ECU. This second phase of the Eureka95 project aimed at the implementation of the first regular wide-screen broadcasting in 1991. (In the same year, Japanese public HDTV broadcasting had already started.) Moreover, Eureka95 participants wanted to achieve HDTV-quality broadcasting of important events (e.g., Olympic games) in 1992 throughout the whole of Europe.

The European Commission wanted to support the application of high-quality television, broadcasting, and satellite technology by developing the HDMAC Directive [8] in May 1992. This Directive aimed to lead Europe to the HDMAC standard via D2MAC. At a later stage, the HDMAC system had to be followed up by a completely digital HDTV system. At that point, however, the U.K. government refused to continue subsidizing the European industry in the development of a European HDTV system. As a result, the HDMAC Directive was abandoned as a policy line [9].

Table 1.4 details the evolution of HDTV systems.

1.6   Digital television

The Federal Communication Commission's (FCC) 1987 initiative concerning an HDTV standard for terrestrial broadcasting resulted in 21 proposals. Most of these proposals were not compatible with the NTSC

Table 1.4
Developments in HDTV Systems

Date	Development
1981	First demonstration of HDTV system by NHK in the United States
1981	Establishment of EBU Working Party V in Europe
1982	Establishment of ATSC in the United States
1983	Establishment of CCIR IWP
1984	MUSE transmission standard by NHK in Japan
1985	CCIR proposal for program production world standard (1,125 lines with 60-Hz interlaced scanning)
1986	Start of Eureka95 project in Europe
1987	Initiative to develop HDTV standard for terrestrial broadcasting in the United States
1991	Regular HDTV broadcasting in Japan
1992	European Union HDMAC Directive
1992	Pilot HDTV broadcasting in Europe

standard and did not meet the HDTV system requirements. In 1992, only four proposals were left. One of them, filed by General Instruments on July 1, 1990, concerned the first proposal for a completely digital HDTV system. However, the FCC mandated that the industry agree on a single proposal. Accordingly, in May 1993, General Instruments and the three other parties that had proposed digital systems—AT&T/Zenith, DSRC/Philips/Thomson, and MIT—formed the *Grand Alliance* (GA), whose objective was to develop an HDTV standard for digital terrestrial broadcasting. The GA adopted the *Motion Pictures Expert Group's* (MPEG's) *MPEG-2* standard for video source coding, system information, and multiplexing and a Dolby standard called *AC-3* for multichannel audio source coding. Additionally, the GA specified a transmission standard for digital terrestrial broadcasting, as well as a standard for transmission via CATV networks. The ATSC plays a role as the keeper of this standard, which is referred to as the *GA HDTV system*.

In reaction to the developments in the United States, the Scandinavian *HD-DIVINE* project to develop an HDTV standard for digital terrestrial broadcasting started in 1991. Moreover, Swedish television launched the idea of a pan-European platform for European broadcasters, with the objective of developing digital terrestrial broadcasting.

Meanwhile, in Germany, conversations took place concerning a feasibility study on current television technologies and the alternatives for the development of television in Europe. Late in 1991, the German government recognized the strategic importance of DTV in Europe and the need for a common approach. Accordingly, the German government invited broadcasters, telecommunication organizations, manufacturers, and regulatory authorities in the field of radio communications to an initial meeting that led to the formation of the *European Launching Group* (ELG) in the spring of 1992. Subsequently, the ELG expanded, and on September 10, 1993, 84 European broadcasters, telecommunication organizations, manufacturers, and regulatory authorities signed a memorandum of understanding forming the *European DVB Project* (DVB) [10, 11].

Meanwhile, however, the European market was demanding more television channels rather than a system with better performance such as HDTV. The application of compression techniques on digital signals allows for a dramatic bandwidth reduction so that more channels can be created within the same available bandwidth. An HDTV signal, however, requires more bandwidth than a normal television signal. This also applies to the digital domain. Moreover, digital transmission allows the application of forward error correction, which results in a better display quality. Hence, DVB is aimed at a normal digital wide-screen (16:9) television, rather than digital HDTV.

DVB decided to adopt the MPEG-2 standard for audio and video source coding, system information, and multiplexing. Additionally, it developed specifications for digital transmission via satellite, CATV, and later terrestrial networks. DVB also specified elements of a European digital *conditional access* (CA) system.

Currently, DVB is specifying transmission systems for the provision of interactive services. According to the DVB transition model, the end user needs a set-top box for the conversion of digital signals into a PAL or SECAM signal. At a later stage, when the signal processing within the television set is also digital, this conversion is no longer required, and a complete DTV system will be achieved.

The European Commission not only supported DVB financially, but, together with the Member States, developed a Directive on television standards [12] as well. The DVB specifications, which were turned into standards by the *European Telecommunications Standards Institute* (ETSI), became mandatory by means of the Directive. The Directive also put an emphasis on the structuring of the market, especially in the field of CA. This contrasted with the HDMAC Directive, which was specifically developed to set a standard [13]. The European Parliament approved the television standards Directive in October, 1995.

In the United States, the GA could not satisfy all needs with the FCC mandate. Consequently, it proposed the development of a digital CATV transmission standard similar to the DVB specifications. Moreover, the *National Association of Broadcasters* (NAB) initiated a feasibility study on the use of the European (draft) digital terrestrial transmission specifications. DirecTV launched the first digital satellite broadcasting in the United States in June, 1994, while in Europe the French Canal Satellite launched the first digital satellite television in April, 1996 [14]. The technology used by DirecTV was developed in cooperation with DVB members parallel to the work on the DVB digital satellite transmission specifications. Hence, the two transmission systems show many similarities.

To fully benefit from the success of the MUSE transmission standard, Japan officially started the development of DTV in the summer of 1994. The Japanese *Ministry of Post and Telecommunications* (MPT) was founded by the Digital Broadcasting Development Office to coordinate the development of DTV. By that time, the European offices of several Japan-based enterprises had already participated in the DVB project. This is probably why Japan adopted the MPEG-2 standard for source coding and system information and why the Japanese proposals for digital transmission systems are similar to the DVB specifications. In October 1996, PerfecTV started the first public digital satellite broadcasting in Japan.

Table 1.5 lists significant events in the evolution of DTV systems.

1.7 Summary and conclusions

The development of television officially started in 1884 when German Paul Gottlieb Nipkow patented his mechanical television system. Several other Europeans and Americans, who constantly improved Nipkow's system, can be considered the first pioneers.

With the introduction of electronic systems, regular television broadcasting began in Europe in 1936, followed by the United States and later

Table 1.5
Developments in Digital Television Systems

Date	Development
1990	First proposal for a completely digital HDTV system for terrestrial broadcasting by GI
1991	Scandinavian HD-DIVINE project on digital HDTV standard for terrestrial broadcasting
1992	Formation of the ELG
1993	Formation of the GA in the United States
1993	Initiation of the European DVB Project
1994	Founding of the Digital Broadcasting Development Office in Japan by MPT
1994	First public digital satellite broadcasting in the United States by DirecTV
1994	European standard on digital *direct-to-home* (DTH) satellite broadcasting by DVB
1994	European standard on digital broadcasting via CATV networks by DVB
1995	European standard on digital *satellite master antenna television* (SMATV) by DVB
1995	ATSC DTV standard A/53 in the United States
1995	European Union television standards Directive
1995	Specification of common scrambling algorithm for CA by DVB
1995	ATSC digital audio compression (AC-3) standard A/52 in the United States
1996	Specification of common interface for CA by DVB
1996	First public digital satellite broadcasting in Europe by Canal Satellite
1996	European standard on digital *multipoint video distribution systems* (MVDS) by DVB
1996	First public digital satellite broadcasting in Japan by PerfecTV
1997	European standard on digital MMDS by DVB
1997	European standard on digital terrestrial broadcasting by DVB

Japan. Depending whether the electrical power was supplied at 50 Hz or 60 Hz, countries around the world used a television system with interlaced scanning at a frequency of either 50 Hz (Europe) or 60 Hz (United States and Japan).

Although the first proposals for color television date from 1904 in Germany, the United States was the first country to provide public color television broadcasting in 1954 using its NTSC standard with 525 lines and 60-Hz interlaced scanning. This standard was adopted by Japan, which started public broadcasting six years later. In Europe, public broadcasting was launched in 1967 using the European SECAM and PAL standards, which were based on 625 lines and 50-Hz interlaced scanning.

In the mid 1960s, an attempt at a high-quality world standard (HDTV) was made in Japan. With the development of the MUSE satellite transmission standard for HDTV with 1,125 lines and 60-Hz interlaced scanning, Japan was far ahead of the rest of the world. Europe reacted, and the HDMAC satellite transmission standard based on 1,250 lines with 50-Hz interlaced scanning was developed. The United States followed a different course with an initiative to develop an HDTV standard for terrestrial transmission.

The United States' initiative resulted in the establishment of the GA, which aimed to develop a completely digital HDTV system for terrestrial broadcasting. In Europe, on the other hand, the market's demand led to the development of a European normal wide-screen (16:9) television system. Within the European DVB, project priority was given to digital transmission via satellite and CATV networks. The terrestrial transmission system was specified later. Several parties in the United States adopted the DVB satellite specifications and will probably adopt the DVB specifications for transmission via CATV networks. In Japan, the DVB satellite, CATV network, and terrestrial transmission specifications will most likely be adopted. Hence, the DVB satellite and CATV specifications may become world standards. In digital terrestrial systems, it seems that there will be two options (DVB or GA).

DTV is more than simply the broadcasting of television programs. With the application of digital techniques, television services can be efficiently provided via several kinds of telecommunication networks. This results in a convergence of the traditional broadcasting and telecommunications sectors, and as these traditional boundaries disappear, new infrastructures for the provision of interactive multimedia services arise. These infrastructures can be regarded as future electronic highways.

References

[1] Robida, A., *Le Vingtième Siècle*, 1883.

[2] Encyclopaedia Brittanica, *Macropaedia, Ready Reference and Index*, Volume 9, 1974, p. 870.

[3] Encyclopaedia Brittanica, *Macropaedia, Knowledge in Depth*, 15th edition, Volume 18, 1983, pp. 105–123.

[4]   Nederlands HDTV Platform, *Handboek High Definition Television, Part I,* Kluwer Technische Boeken B.V., Deventer, December 1992.

[5]   Sarlemijn, A., and M. De Vries, *The Piecemeal Rationality of Application Oriented Research:An Analysis of the R&D History Leading to the Invention of the Plumbicon in the Philips Research Laboratories,* Kluwer Academic Publishers, 1992.

[6]   Nederlands HDTV Platform, *Handboek High Definition Television, Part VII,* Kluwer Technische Boeken B.V., Deventer, December 1993.

[7]   Reimers, U., *Digitale Fernsehtechnik, Datenkompression und Übertragung für DVB,* Springer, April 1995.

[8]   Directive 92/38/EEG of the European Council of 11 May 1992 on the establishment of standards for the satellite transmission of television signals, PbEG L137.

[9]   Smits, J., *DVB: fundament op weg naar Europese Elektronische Snelweg?,* Kabeljaarboek, December 1994.

[10] Reimers, U., *European Perspectives on Digital Television Broadcasting—Conclusions of the Working Group on Digital Television Broadcasting (WGDTB),* EBU Technical Review, No. 256, Summer 1993, pp. 3–8.

[11] DVB Project Office, *DVB Blue Brochure,* 2nd Edition, 19 May 1995.

[12] Directive 95/47/EC of the European Parliament and of the Council of 24 October 1995 on the use of standards for the transmission of television signals, O.J. L281/51, 23 November 1995.

[13] de Bruin, R., *Technologie Beleidsonderzoek naar Interactieve Digitale Video-diensten met Conditional Access,* Technische Universiteit Eindhoven, October 1995.

[14] Moroney, J., and Th. Blonz, "Digital Television: The Competitive Challenge for Broadcasting and Content", *Ovum Reports,* 1997.

Contents

Theoretical framework

2.1 Introduction

This chapter provides a theoretical framework for DTV. Part of this framework consists of a model that is used to describe the different services that can be provided via DTV systems. The advantage of these systems is their ability to provide large-scale interactive services, instead of providing only traditional distribution services. Depending on the content's economic value, some of these services may be provided via a CA system.

The second part of the framework is formed by a policy model. As discussed in Section 2.3, this model is a useful tool for making a functional distinction on which several types of policies can be based. Finally, the services model and the policy model are combined to provide an overview of the theoretical framework, which also includes the services' various information streams.

2.2 Services

This section first discusses interactive services and CA services. Next, a generic services model concerning a categorization of different types of services is presented. Finally, examples are used to illustrate the services model.

2.2.1 Interactive services

The traditional principle of television is that the broadcaster's content is distributed via a broadcast network to the end user. With respect to these kinds of services, television can be considered a passive medium. As concluded in Chapter 1, DTV enables more than the distribution of content only. It allows a large-scale provision of interactive multimedia services via the television medium. This implies that in communication, the end user is able to control and influence the subjects of communication, with the control and influence taking place via an interactive network [1]. Hence, the user is able to play a more active role than before.

2.2.2 Conditional access services

(Interactive) television services can be provided via CA systems. In this context, a CA system ensures that only authorized users (i.e., users with a valid contract) can watch a particular programming package [2]. In technical terms, a TV program is broadcast in encrypted form and can only be decrypted by means of a set-top box. The set-top box incorporates the necessary hardware, software, and interfaces to select, receive, and decrypt the programs. Chapter 11 discusses the aspects of CA in more detail.

2.2.3 Services model

Depending on the different forms of communication and their application, two categories of telecommunications services can be distinguished: interactive services and distribution services (see Figure 2.1). These categories can be further divided into several subcategories: The interactive services are divided into registration, conversational, messaging, and retrieval services, while the distribution services, in turn, are divided into services with and without individual user presentation control. This

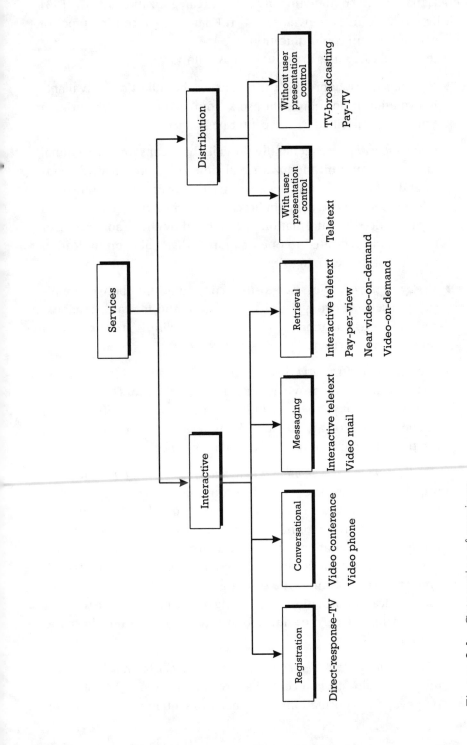

Figure 2.1 Categories of services.

model is based on a combination of a services categorization by CCITT [3] and the model from Bordewijk and Van Kaam [4] concerning types of information flows in communication.

The several categories are described as follows:

> ▶ *Registration services* provide the means for the collection of available information from individual users (sources) by a center during a time that is specified by the center per subject.

> ▶ *Conversational services* generally provide the means for bidirectional dialogue communication with real-time (no store-forward) end-to-end information transfer from user to user. The flow of user information may be either bidirectional symmetric or bidirectional asymmetric. The information is generated by the sending user or users and is dedicated to one or more individual communication partners at the receiving site.

> ▶ *Messaging services* offer user-to-user communication between individual users via storage units with store-and-forward, mailbox, and/or message handling (e.g., information editing, processing, and conversion) functions.

> ▶ The user of *retrieval services* can retrieve information stored in information centers and provided for general public use. This information will be sent to the user only upon demand. The information can be retrieved on an individual basis. Moreover, the user controls the time at which an information sequence begins.

> ▶ *Distribution services without user individual presentation control* include broadcast services. They provide a continuous flow of information that is distributed from a central source to an unlimited number of authorized receivers connected to the network. The user can access this flow of information *without* the ability to determine the instant at which the distribution of a string of information will be started. The user cannot control the start and order of the presentation of the broadcast information. Depending on the point of time the user accesses the service, the information may not be presented from the beginning.

> ▶ *Distribution services with user individual presentation control* also distribute information from a central source to a large number of users. However, the information is provided as a sequence of information

entities (e.g., frames) with cyclical repetition. As a result, the user has the ability to individually access the cyclical distributed information and can control the start and order of presentation. Due to the cyclical repetition, the information entities selected by the user will always be presented from their beginning.

Figure 2.1 also contains several examples of types of services. These services can be provided with the application of digital techniques and, depending on the content, may or may not be provided by way of a CA system. Some of the types of services illustrated by Figure 2.1 are explained as follows.

- *Direct-response-TV* implies that the user can respond directly to the provided program. Examples are interactive TV quiz shows or interactive commercials. The required return channel can, for example, be realized via a public telephone network or a bidirectional CATV network.

- In the case of *pay-TV*, a CA system is used to allow only authorized users to watch a particular programming package. The programming package is broadcast in encrypted form and can only be decrypted by means of a set-top box.

- A *pay-per-view* (PPV) system basically uses the same technique as pay TV. The only difference is that the user now pays per program, rather than paying for the entire programming package. Technically, the system is extended with an ordering system. The required return channel can be realized via the same networks as mentioned above.

- A *video-on-demand* system enables an individual user to demand a program, which is stored in a database, at a time specified by this user. Moreover, the user may have the ability to perform such functions as stop, forward, or play back the selected program.

- *Near-video-on-demand* refers to a system that starts the same program on a different channel with a pause (e.g., every 10 minutes). This requires the use of a considerable number of channels. These channels can be created through the application of digital compression techniques, by which means the required bandwidth per channel decreases dramatically. In contrast with the video-on-demand system, the user has to wait a short time in order to watch a selected program. Hence, the near-video-on-demand system *nearly* provides the same convenience as the video-on-demand system.

▶ With linear teletext, several teletext pages, embedded in the TV signal, are broadcasted to the TV set via a transmission medium. The selected teletext page is stored in memory, after which it can be watched. In case of *interactive teletext* a teletext page is transmitted to an individual user or a user group. The request for the page concerned can be processed via a return channel through, for example, a public telephony network. By means of this interactive teletext system messaging services (e.g., personal insurance information) can be supported. Moreover, the same technique allows the retrieval of data from external databases. In this case, one can, for example, retrieve travel information.

2.3   Policy

This section discusses the layer model—which can be used to make a functional distinction on which several types of policies can be based—and uses examples to illustrate the layer model's scope and application. In addtion, the layer model is combined with the services model, thereby allowing an overview of the theoretical framework. This overview also discusses the services' information streams.

2.3.1   The layer model

In telecommunications and broadcasting, the central element is the content, rather than the way of transporting it. Transport can be established in several ways, for example, via electromagnetic (cable and ether) or optic (optical fiber) transmission. In addition to the difference between content and transport, terminal equipment can be distinguished. Terminal equipment is needed for the presentation of content. The user, in turn, may use the same or other terminal equipment to send data back to the source. Hence, interactive communication is achieved (see Figure 2.2).

Within content and transport a further distinction can be made. Content can be divided into information and information services. In turn,

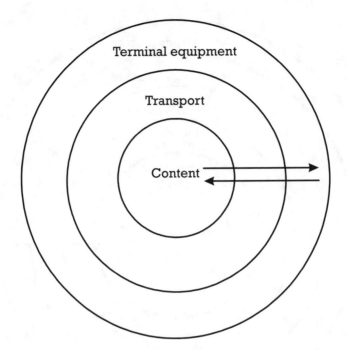

Figure 2.2 Difference between content, transport, and terminal equipment.

transport concerns value-added services, network services, and transmission capacity (see Figure 2.3).

The terms used in Figure 2.3 are defined as follows:

▶ *Information* is meaningful data that is transported in a particular way.

▶ When the combination and provision of information takes place in a more or less institutionalized way, one speaks of an *information service*.

▶ A *value-added service* is a service that, with the use of the routing capacity of one or more network services, establishes an additional function to the network service(s).

▶ A *network service* is a service that, with the use of the transmission capacity provided by one or more telecommunication infrastructures, provides routing for the end users.

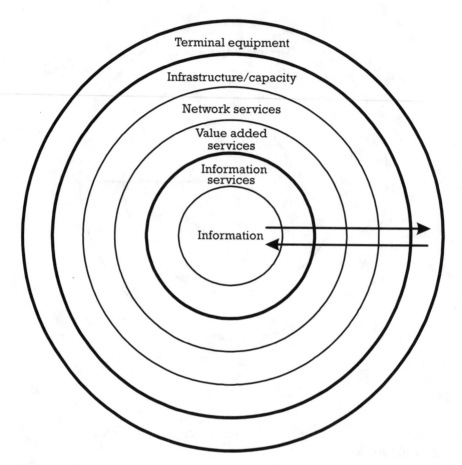

Figure 2.3 Functional distinctions in telecommunications and broadcasting.

- The *telecommunication infrastructure* is the transmission capacity that can be used for the transport of signals between defined network termination points.

- *Terminal equipment* is a construction or a combination of constructions that is meant to be directly connected to a public telecommunications network via a network termination point.

- A *network termination point* is the whole of physical connections with their technical specifications, which are part of a telecommunications network and are needed to obtain access to this network and to efficiently communicate via this network.

It is worth noting that the definition of value-added services or value-added networks varies from country to country but is generally worded to cover any service falling outside the definition of a basic network service. This definition allows entrepreneurial companies to establish services at a premium price, based upon network and transmission equipment leased from the public telephone operator [5].

Terminal equipment can be a telephone, telex, fax, or modem but can also refer to complicated equipment and private networks. The latter forms the link between the user's own facilities and the public telecommunications network.

The functional distinction just made is presented in a different way in Figure 2.4. Because of its layered construction, this model is called the *layer model.*[1]

Within each layer, the model includes several examples. The examples concerning the network services and transmission capacity layers are more specific. A network service is constructed for a specific application. Each application requires a certain bandwidth in the frequency domain. Depending on the application, each network incorporates a specific infrastructure's transmission capacity. For example, a radio and television network service makes use of a CATV infrastructure's capacity for cable transmission. However, a different radio and television network service may use the radio spectrum's capacity for satellite broadcasting. With the application of compression techniques, this network service can even be provided via the PTT (*post telegraph and telephone*) infrastructure's capacity. On the other hand, an infrastructure's capacity may be used for more than one application. For example, a CATV infrastructure can provide capacity for a fixed telephony network service as well as a radio and television network service.

In principle, all network services may use different types of infrastructures' capacities, and an infrastructure's capacity may be used for one or more network services. Hence, a matrix of all different kinds of possibilities arises. In some cases, however, legal and/or technological barriers still exist. For example, CATV network operators are not always allowed to provide telephony services, and not all CATV networks are yet capable of supporting two-way communication. The removal of legal barriers is required to stimulate this technological innovation.

---

1. See also [6, 7] for layer modeling in telecommunications.

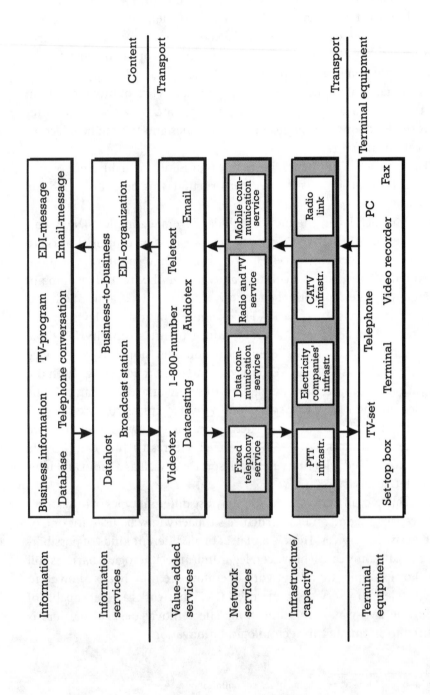

Figure 2.4 The layer model.

The next layer concerns terminal equipment, which can be connected to the network. On the dividing line between transport (i.e., the infrastructure's capacity) and terminal equipment, the network termination point is defined. The equipment needed for a service's realization has to be regarded as part of the service and not as terminal equipment. This equipment is referred to as supporting equipment. From a legal perspective, terminal equipment is directly connected to the network termination point.

Finally, it has to be stated that it is not necessary to make use of an information service and/or a value-added service. For example, a telephone conversation does not involve an information service and does not necessarily require a value-added service.

2.3.2 Scope and application of the layer model

The layer model describes a functional separation between content, transport, and terminal equipment. This model can be applied in several ways. First, it can be used to make a distinction between certain legal or policy domains. Some legal or policy frameworks are oriented on content (e.g., media policies), others on transport (e.g., telecommunications and broadcasting policies) and/or terminal equipment (e.g., telecommunications and consumer equipment policies). Next, the various activities of the actors involved can be allocated in the model's different layers. Furthermore, from an economic perspective this model can be regarded as a value-added chain. From the information layer to the terminal equipment layer, economic value is added to the information product. Finally, this model makes a clear distinction between services on one hand and the infrastructure's capacity on the other hand. This allows a matrix of network services and infrastructure's capacities.

The layer model does not make a distinction from a purely technical perspective. In contrast, the *open systems interconnection* (OSI) model and the four-layer model [8] are models that describe several layers from this limited perspective. The OSI model's layers are the *physical, data link, network, transport, session, presentation,* and *application* layers. The four-layer model describes the *connections, transport functions, telematic functions,* and *additional norms* layers. It is beyond the scope of this section to discuss both technical models in more detail. Because of the layered structure of all three models, one might believe their application is similar. However, the layer model proves its value because of its broader scope.

2.3.3 Services and information streams

Section 2.2 discusses several categories of interactive and distribution services. These services and their information streams can be visualized within the layer model (see Figure 2.5).

To understand Figure 2.5, consider, for example, an interactive DTV program in which the viewer of an international tennis match has the ability to choose the camera position. From the perspective of the layer model, the images which are produced by several cameras, and the accompanying sounds form the information. A broadcaster combines this information with the voice of a local or regional reporter and includes this program in its programming package. Moreover, the broadcaster encrypts this programming package so that only authorized viewers with whom the company has a valid contract can watch it. Hence, the broadcaster provides an information service and a value-added service respectively. Next, the programming package, including the international tennis match, is provided to the user via a bidirectional radio and television network service, which, in turn, uses a CATV infrastructure's capacity. Alternatively, the radio spectrum's capacity could be used for satellite broadcasting. Finally, the international tennis match is provided to the user's set-top box after which decryption takes place and the program can be watched. With respect to the services model, the broadcaster provides a distribution service without individual user presentation control.

However, the viewer is capable of choosing the camera position. Hence, the viewer becomes the stage manager. The user's starting point is the terminal equipment layer within the layer model. By means of an interactive set-top box, the viewer sends a signal back to the broadcaster, and this signal indicates the camera position. This signal can be transported via the bidirectional radio and television network, which uses the CATV infrastructure's capacity. In the case of satellite broadcasting, the viewer can send his or her signal via a fixed telephony network, which uses the PTT infrastructure's capacity. The path along which the viewer sends his signal to the broadcaster is called the return channel. Finally, the broadcaster provides the required information (i.e., the

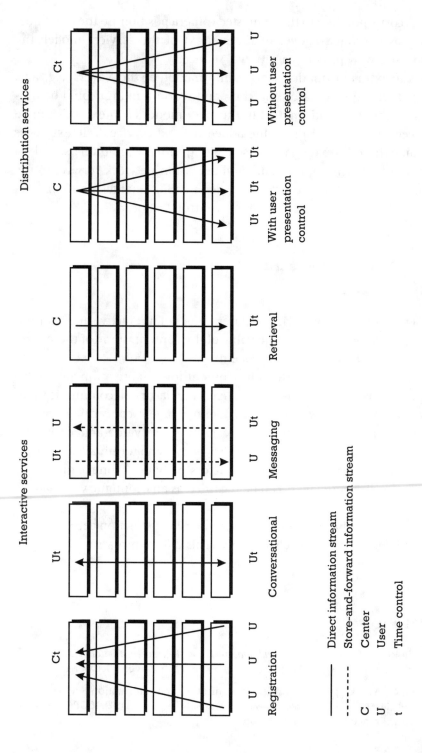

Figure 2.5 Services and information streams within the layer model.

images corresponding to the requested camera position) to the individual viewer as described above. In the context of the services model, the broadcaster now provides a retrieval service.

As an extra information service, information on the players (e.g., age, nationality, results of latest matches, and world ranking) could be provided during the tennis match. In the layer model's context, this can be achieved by means of the value-added teletext service. Teletext pages including this information can be sent along with the program's signal via the routing capacity of the radio and television network service. In the services model, the broadcaster provides an additional distribution service with individual user presentation control.

2.4 Summary and conclusions

The services model categorizes different types of interactive and distribution services. Depending on the value of the content, some of these services may be provided via a CA system.

The layer model is a useful instrument for making a functional distinction on which several policies (e.g., media or telecommunications policies) can be based. As such, it defines layers concerning content (information and information services), transport (value-added services, network services, and infrastructure/capacity), and terminal equipment. The model can be used to allocate the actors' activities in one or more of its layers as well. These layers can be regarded as an economic value-added chain.

In addition, the layer model provides a clear distinction between services and the infrastructure's capacity. This allows a matrix of network services and infrastructure capacities.

References

[1]   de Bruin, R., *Technologie Beleidsonderzoek naar Interactieve Digitale Video-diensten met Conditional Access*, Technische Universiteit Eindhoven, October 1995.

[2]   OECD Working Party on Telecommunication and Information Services Policies, *Conditional Access Systems: Implications for Access*, DSTI/ICCP/TISP(97)7, Paris,France,September 1997, pp. 15–16.

[3]   CCITT, *Blue Book Volume III —Fascicle III.7, ISDN general structure and service capabilities, Recommendations I.110 - I.257*, Geneva, 1989.

[4]   Bordewijk, J. L., and B. van Kaam, *Allocutie*, Baarn, 1982.

[5]   Clark, M. P., *Networks and Telecommunications, Design and Operation*, John Wiley & Sons, 1991.

[6]   Bekkers, Rudi and Jan Smits, *Mobile Communications: Standards, Regulation and Applications*, Boston: Artech House, 1990.

[7]   Smits, J. and J. deVries, *Het lagenmodel een toekomstbvaste basis voor inrichting en regulering van de telecommunicatie*markt?, Informatie en Informatiebeleid, Winter, 1993.

[8]   Ministerie van Binnenlandse Zaken, *Telematica-atlas, Openbare Sector*, Samsom, November, 1993.

Contents

Technological and market convergence

3.1 Introduction

The technological developments in information and communications technologies have led to a convergence of speech/audio, data, text, graphics, and video and thus to multimedia applications. At the same time, people in modern society are becoming more and more individualistic. As a result, service providers are forced to focus on individual consumer demands. Since interactivity allows them to obtain individual feedback from their customers, technological convergence and social individualization lead to the development of interactive multimedia services.

The technological convergence also affects several traditional service-oriented sectors of the communications industry, namely the entertainment, information, telecommunications, and transaction sectors: As their products are integrated, these sectors are becoming more and more dependent on each other. In Chapter 2, the layer model is used to identify the several actors' activities in these sectors. This chapter describes the changes in their activities within, as well as between, sectors and explains the aspect of market power in the economic value-added chain.

3.2 Convergence among traditional sectors

Four traditional sectors that play an important part in the development of interactive multimedia services in the context of DTV are the entertainment, information, telecommunications, and transaction sectors. The entertainment sector's services are aimed at individual consumers in their home environment. Examples are television, pay TV, and electronic games. The information sector provides services that normally require the user to take the initiative, and the user, rather than the service provider, selects all information. This entails, for example, online/off-line services and database retrieval. It should be noted that the written press (e.g., newspaper publishers) is also part of this sector. However, it is beyond the scope of this book to discuss these actor's services.

The telecommunications sector provides services for communication between persons and/or systems. Such communication takes place via a telecommunications network.

Finally, the transaction sector provides services that concern the transfer of information in the form of an individual participant's instruction to the center at a time specified by the individual participant, with the objective of accomplishing an agreement. Transactions can be implicitly included in a service. For example, in the case of PPV, the ordering of a movie automatically leads to payment. On the other hand, transactions can be explicit. In this case, the transaction service supports another service. Examples are teleshopping, electronic banking, and electronically ordering cinema tickets. These services do not necessarily lead to a transaction, since these services can also be used to retrieve product or credit/debit balance information.

The entertainment sector's and information sector's products cannot always be easily distinguished. Information that is simply entertaining for one user can provide much needed facts for another user. A basketball game, for example, can be considered entertainment. However, the same game can contain necessary information for players and/or coaches analyzing techniques and tactics. As a result of the convergence of the information and entertainment sectors, the term *infotainment* is often used. Infotainment services like pay-TV or online services are often provided via telecommunications networks. This implies a convergence between the infotainment and telecommunications sectors. This convergence is also evident in the existence of electronic games that can be played by several people at once using telecommunications networks. Convergence between the infotainment and transaction sector occurs when users make explicit transactions for the payment of infotainment products. For example, a user can order a movie via a video-on-demand service and pay for it with a prepaid smart card. The required card reader can be incorporated in a set-top box.

3.3 Layer modeling of sectors and actors

Chapter 2 explained that the layer model can, among other functions, be used to allocate the actors' activities. For the purpose of this chapter, the layer model is used to construct a matrix of layers and the traditional sectors (see Figure 3.1). The matrix shows that several actors are active in more than one layer within a sector as well as in more than one sector. It has to be stated that, depending on the layer and the sector, the number of actors is not always equal in all parts of the matrix.

The group of actors that is involved in information production consists of rightful claimants of special information (e.g., films, articles, databases, or video games). In transactions, it is difficult to address the information's rightful claimant. In this case, the financial institutions (i.e., the initiators of this information) are considered to be the information producers [1]. The information service providers can be broadcasters, packagers, online service providers, and commercial and financial service organizations. Value-added service providers, in turn, provide subscriber management, orders, billing, and other additional network tasks for the exploitation of information services. The providers of network services are public telecommunications organizations and CATV, terrestrial, and satellite operators. The last group is mainly active in the field of *direct broadcast satellite* (DBS) services.

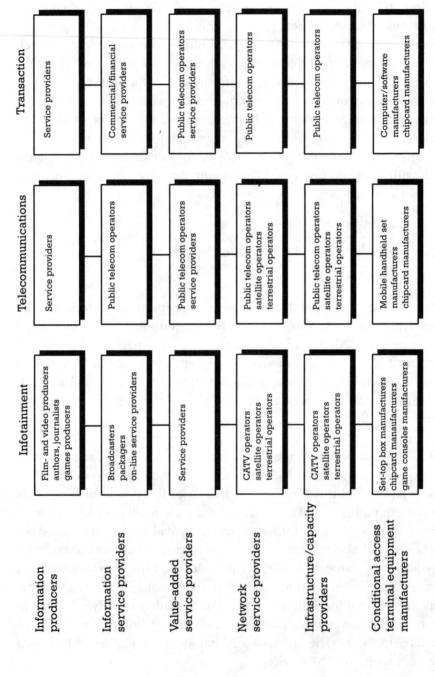

Figure 3.1 Layer modeling of sectors and actors.

The infrastructure/capacity providers' layer involves, more or less, the same actors as those participating in the provision of network services. In some cases, this has a historical background. For provision of the public telephony and radio and television services, the PTT and the CATV operators traditionally had to take care of the network services and the infrastructure's capacity. However, these actors do not necessarily have to be active in both layers at the same time. For example, some public telecommunications organizations specialize in the laying of transatlantic cables. Network service providers, in turn, can lease part of this cable's transmission capacity for their networks. Another example is a satellite operator that has to obtain a license for the use of a part of the radio spectrum in order to provide its services.

Finally, there are the manufacturers of CA terminal equipment (e.g., set-top boxes and cash machines). These actors must not be automatically identified as *gatekeepers*. Gatekeepers exploit the CA system. Often, an information service provider (e.g., a pay-TV operator) exploits this system. However, CA terminal equipment is an important instrument in executing the gatekeeper's function.

3.4   Changes in actors' activities

Currently, several actors in the provision of information and communication are changing their activities. In most cases, this concerns an extension of the core business toward other layers within the economic value-added chain or even toward other sectors. This section explains the changes in actors' activities, as well as the character of these changes, and discusses the actors' motivation for these changes.

3.4.1   Market behavior

The actors are allocated in the economic value-added chains of the traditional sectors of infotainment, telecommunications, and transaction through the matrix in Figure 3.1. The character of changes in the actors' activities, in the case of the extension of activities, can be categorized by means of several types of integration, defined as follows:

▶ *Horizontal integration:* The extension of activities toward two or more sectors within the same layer of the economic value-added chain;

▶ *Vertical integration:* The extension of activities toward two or more layers within the same sector's economic value-added chain;

▶ *Diagonal integration*: The extension of activities toward two or more sectors and not within the same layer(s) of the economic value-added chain(s).

Concerning these types of integration, a further distinction in actual market behavior can be made. For this purpose, the following actual behaviors are defined [2]:

▶ *Competition* is the rival behavior of market parties, which act autonomously toward the achievement of a certain objective. The behavior is subject to uncertainty. Competition can lead to a market party's expansion. When a market party expands without cooperating with, or taking over, rivals, the market party achieves internal growth.

▶ *Concentration* is the expansion of a market party through fusion, takeover, or the obtaining of control through ownership and management, in the case of a minority participation.

▶ *Cooperation* occurs when independent market parties strive toward a common objective through united means or behaviors. Cooperation between market parties can exist in binding agreements, mutually adjusted actual behavior, or the foundation of joint ventures.

Figure 3.2 presents all possible types of market behavior within the economic value-added chain(s).

Market and literature research [3] on the infotainment sector has shown that three important trends can be distinguished within this sector:

1. Horizontal integration based on competition (in the layers concerning terminal equipment, infrastructure/capacity and network services);

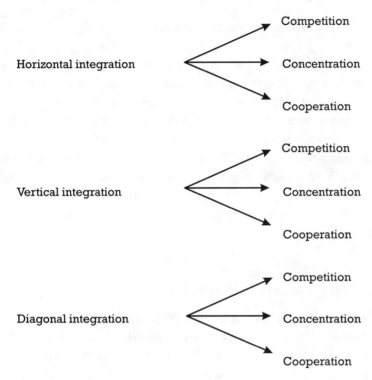

Figure 3.2 Types of market behavior.

2. Vertical integration based on cooperation (in all layers, except the information layer);

3. Vertical integration based on competition (in all layers, except the layers concerning information and information services).

Sections 3.4.2–3.4.5 will discuss these trends for each layer's actors in more detail.

3.4.2 Conditional access terminal equipment manufacturers

The horizontal integration in the CA terminal equipment manufacturers' layer mainly occurs among chip card manufacturers, which operate on the market in all three sectors on the basis of full competition. Chip cards are becoming more and more standardized. As a result, the concerned terminal equipment incorporates standard solutions for the application of chip cards.

When set-top box manufacturers integrate vertically, they integrate via cooperation. This cooperation is mainly embodied by joint ventures with network service providers, value-added service providers, and/or information service providers. At present, game console manufactures do not show any sign of horizontal or vertical integration. Their only relation is with the information layer's game producers. It is very well possible that in the near future interactive games will be played on the PC or television with several players via telecommunications networks. In this case, cooperation with the actors from the intermediate layers is likely to occur.

3.4.3 Network service providers

Network service providers often manage the network's infrastructure capacity as well as the network service itself. Hence, both layers are vertically integrated to a large extent. Within the layers concerning network services and infrastructure/capacity, horizontal integration based on competition takes place. Network services are being provided more and more outside the traditional sector. For example, CATV operators have started to provide public telephony services. Hence, they now operate in the traditional telecommunications sector. In turn, public telecommunications organizations are conducting research on the provision of high-quality video images via their telephony networks. Similarly, the transaction sector has always made use of the public telecommunications organizations' leased lines for their financial transactions. In principle, CATV operators have the potential to do the same.

In addition to the provision of radio and television network services, network service providers (mainly CATV operators) aim to provide (their own) value-added services. In the value-added services layer, the financial margins are higher. Network providers offering value-added services are able to improve their profile with customers. This vertical integration process has a competitive character.

On the other hand, network service providers also vertically integrate on the basis of cooperation by forming joint ventures with information service providers and CA terminal equipment (i.e., set-top box) manufacturers. This allows network service providers to share in their competitors' profits but may sometimes introduce a conflict regarding which party should act as the gatekeeper. This is because more and more network service providers, such as satellite and CATV operators, are providing CA services as a value-added service. For this purpose, they are allocating (proprietary) set-top boxes to their customers' premises. Their motivation is that they want to become gatekeeper for, at the very least, their own services, rather than providing only their infrastructures' capacity. Moreover, the role of gatekeeper allows them to increase their profile and to build up customer relationships. Hence, the network service providers also vertically integrate on the basis of competition.

3.4.4 Value-added service providers

Value-added services provide an additional function to the basic network service. When network services are used in sectors other than the traditional sector, value-added services will also be introduced in these sectors. In general, value-added service providers do not undertake any large-scale initiatives to any form of integration. At this moment, the value-added service provider layer features none of the horizontal integration that takes place in the underlying layers. However, there are some interesting applications that show signs of diagonal integration.

For example, public telephony networks are used to provide return channels to facilitate interactive television services. The value-added 1-900 service allows the customer to order a specific information service. It can also be used to generate a cash flow. In the infotainment sector, meanwhile, a value-added transaction service called *E-cash* has been introduced. E-cash (an electronic equivalent of cash) can, among other things, be used to pay for information services. The basis of E-cash is the application of cryptography, by which a cyberbank processes payments securely and anonymously via normal telephone lines. Presently, banks are too conservative to do business via unsecured telephone lines, but, on the other hand, they see many potential customers. Thus, in the future they may act as Cyberbanks as well. In the United States, experiments with E-cash have been done on the Internet.

3.4.5 Information service providers

Many pay-TV operators act as information service providers (i.e., programming packagers) and value-added service providers (i.e., CA services). These operators are vertically integrated to a large extent with satellite operators and set-top box equipment manufacturers on the basis of cooperation. In countries with a high cable penetration, pay-TV operators are vertically integrating on the basis of cooperation as well. In these cases, pay-TV operators mainly start joint ventures with CATV operators and set-top box manufacturers.

At this moment, horizontal integration in the information service layer hardly takes place. The information service providers' products still show too many differences. However, as a result of technological developments (i.e., interactive multimedia services), new possibilities are arising for the different sectors' information service providers to combine their products. For example, teleshopping programs entail horizontal integration of the infotainment and transaction sectors.

3.4.6   Information producers

The integration process is developing gradually within the information production layer. In the case of electronic publication of newspapers and/or electronic magazines, publishers and online service providers are also acting as information service providers and sometimes even provide value-added services. Hence, these actors vertically integrate on the basis of competition.

Furthermore, several commercial broadcasters work together with publishers so that they are not fully dependent on the rightful claimants of information. Their market behavior can be characterized as horizontal integration on the basis of cooperation and sometimes even concentration.

An overview of the actors' changes in activities can be presented via the layer modeling of sectors and actors (see Figure 3.3).

3.5   Power in the value-added chain

Within the infotainment sector, information service providers, as well as network providers and CA terminal equipment manufacturers, want to control the access to services. In other words, they all want to function as gatekeeper. Information service providers want to establish a relationship

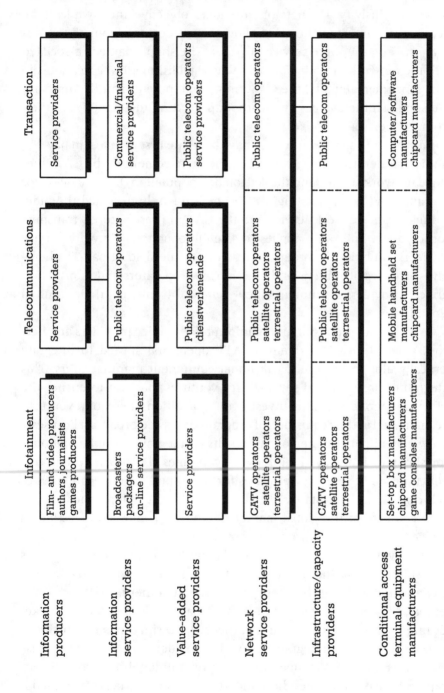

Figure 3.3 Layer modeling of actors' changes in activities.

between consumers and their specific product(s). In doing so, they aim towards exclusivity. The network providers' objective, meanwhile, is the creation of a one-stop shop that allows consumers to obtain all services from one and the same service provider. The shop's name is associated with the network provider, allowing the establishment of a relationship with consumers. Finally, the CA terminal equipment manufacturers try to distinguish themselves by incorporating extra functionalities in their products (i.e., set-top boxes).

Information service providers have been the first to invest in pay-TV. In this manner, they have controlled the management of the CA value-added network service and the set-top box population. In the case of, for example, CATV broadcasting, CATV operators only provided the radio and television network service and the CATV infrastructure's transmission capacity. The same applies to satellite operators. Network service providers' ambition is to deploy activities in the field of pay-TV themselves. This results in tension between CATV operators and information service providers, although they remain aware of their mutual dependence.

With this awareness in mind, information service providers and network service providers (mainly CATV operators) started to integrate vertically along with the CA terminal equipment manufacturers. This integration has spawned several joint ventures that aim to ensure the achievement of a return on investment. In addition, these joint ventures are interested in the acquisition of program rights and broadcast licenses.

In general, one of the positive results of integration can be that one-stop shopping is achieved. Moreover, economies of scale can lead to cost reductions and thus to lower prices for consumers. These economies of scale can also result in standardized solutions for CA systems and the ability to compete in the global market. Moreover, combined investments can lead to a general improvement of service.

However, integration can also lead to undesirable situations: There is a considerable risk that the participating network service provider will offer a better deal to its business partner than to other its partner's competitors. Worse yet, potential competitors may not even enter the market because of this risk. This applies particularly to situations in which the concerned vertically integrated network provider has a monopoly position; for example, CATV operators are often monopolists in a local geographical area. Companies' hesitation to enter the market is increased if the market is only big enough to accommodate a limited number of service providers and if those service providers already have a strong market position. Such a position is often protected by the application of proprietary solutions for CA systems.

3.6 Summary and conclusions

As a result of the convergence of various information and communication technologies, the DTV market is also subject to a convergence process. The actors from the traditional infotainment, telecommunications, and transaction sectors are developing activities beyond the scope of their core business. In the context of DTV, several actors from different layers within the infotainment sector compete to play the gatekeeper's role. At the same time, however, they also integrate vertically on the basis of cooperation: By launching joint ventures, they try to eliminate uncertainties in the achievement of a return on investment, which are characteristic of many emerging markets. Integration can have positive as well as negative effects. The objective of government policies should be to create an open market structure without affecting the incentive to (further) invest in this turbulent market.

References

[1] Schrijver, F. J., O. Gorter, G. Schijns, and R. Keizers, *Electronic Highways: Toegang tot telecom*, Intercai Nederland B.V., 13 June 1994.

[2] Mediaraad, *Advies inzake herstructurering beleid informatievoorziening deel I: Het informatietransport*, 11 June 1993.

[3] de Bruin, R., *Technologie Beleidsonderzoek naar Interactieve Digitale Video-diensten met Conditional Access*, Technische Universiteit Eindhoven, October, 1995.

Contents

United States

4.1 Introduction

The current NTSC standards were created in the 1940s through a joint effort of industry and governmental agencies. However, as is the case with the new DTV standards, a heavy battle between diverse interest groups predefined the outcome. The battle was about whether television should be seen as "viewable radio" or whether watching television should be the equivalent of "enjoying a movie in the living room." A forceful intervention of the *Radio Corporation of America* (RCA) ended the battle. The result was the adoption of the NTSC system as a series of technical standards for television, thus making the radio broadcasting and manufacturing industries the winners. If this outcome had not been achieved, television might have become "theater TV" or "subscription TV" linked by cable.

The new DTV standards were to be created by an *Advisory Committee on Advanced Television Service* (ACATS). This committee was established in November 1987 by the FCC to

49

assist the agency in the establishment of new video standards for the United States.[1] Initially, approximately 23 advanced television proposals—all featuring analog transmission—were presented to the committee. Through proponent mergers and attrition this number was soon reduced to a handful. Subsequently, the *Advanced Television Test Center* (ATTC) was established in June 1988. The ATTC was charged with testing the various advanced television systems in the field and under laboratory conditions. Another important event was the March 1990 FCC announcement for preference of simulcast broadcasting. By showing this preference, the FCC challenged the contenders to deliver HDTV in a single 6-MHz broadcast channel, the same capacity assigned to the NTSC broadcasts. In June 1990, the General Instrument Corporation modified its proposal to fully incorporate digital transmission. Three of the four remaining HDTV systems quickly adopted this technological advance with only the Japanese NHK proposal retaining its original analog transmission scheme. In February 1993, the advisory committee approved the release of the report on testing and data analysis of the five HDTV systems. Based on these tests, the committee decided that the four digital systems had spectrum utilization characteristics far superior to the NHK proposal, which was thereafter eliminated. The advisory committee provided the proponents with a critical choice: to undergo a second (and expensive) round of testing focusing on technical improvements that each system had proposed, or to merge their efforts in a single unified system.

Mid-1993 saw the formation of the GA , whose members were AT&T, General Instrument Corporation, Massachusetts Institute of Technology, Philips Electronics North America Corporation, David Sarnoff Research Center, Thomson Consumer Electronics, and Zenith Electronics Corporation. In 1994, the first GA system was constructed. AT&T and General Instrument jointly built the video encoder. Philips constructed the video decoder. Sarnoff and Thomson cooperated in building the transport subsystem, and Zenith built the modulation subsystem. In March 1995, after integrating the different elements into a complete prototype, the GA HDTV standard was completed. The system was delivered to the FCC for testing, at the ATTC in Alexandria, Virginia, followed by field testing at Charlotte. The proposed standard was submitted to the FCC

---

1. Notice of Inquiry in the Matter of Advanced Television Systems and Their Impact on the Existing Broadcast Service, 2 FCC Rcd 5125 (1987).

for final certification. The proposed HDTV standard was approved by the FCC-ACATS.[2] Finally, congressional approval was granted for the new GA-HDTV standard for the United States.

Trying to formulate conclusions on the standard-setting process for DTV, it is possible to state that the winners of the new digital standards are the representatives of the telecommunications and IT (equipment) industry rather than the broadcasting industry, that won the battle for the DTV standards [1, 2].

4.2 General policy and regulatory environment

Before zooming in on the DTV context, it is necessary to describe the general policy and regulatory environment. This concerns telecommunications as well as the broadcasting policy and regulatory frameworks [3, 4].

4.2.1 The communications act

By 1996, the Communications Act of 1934, after 20 years of fundamental criticism, needed to be updated. After 62 years of enforcing the act, there were so many statutory and court-ordered barriers against competition between segments of the telecommunications industry that renewal became a necessity. These barriers between different operators needed to be eliminated to enable *Bell operating companies* (BOCs), long-distance carriers (*interexchange carriers* [IXCs]), cable companies, broadcasters, and others to compete with one another. The result was the 1996 Telecommunications Act (Pub. L. No. 104-104, 110 Stat. 56, approved February 8, 1996).

The preamble to the 1996 act states that it intends "to promote competition and reduce regulation in order to secure lower prices and high-quality services for American telecommunications consumers and encourage the rapid deployment of new telecommunications technologies." The act promotes direct competition between all telecommunications providers, including terrestrial broadcasters, DBS, mobile

---

2. Fifth Further Notice of Proposed Rulemaking, adopted May 9, 1996, FCC 96-207 (20 May 1996), regarding the Advisory Committee recommendations for a technical standard for digital broadcast.

communication services, cable providers, the BOCs, and long-distance telephone companies. The next decade will see a rapid rise in the production and DTH delivery of video programming. This will not be achieved by traditional programmers and distributors but by corporate organizations, advertisers, direct marketers, public relations firms, and even political strategists. They are the interested parties that want to reach directly into domestic and international markets. Mergers, acquisitions, and alliances across telecommunications, computer, and traditional media lines could encourage cartel-like tendencies in television (and interactive) programming. It is probably safe to predict that, as a consequence of increased channels of distribution and an expanding production community, television programming will increasingly come via "full-service" communications providers.[3] Increased channel capacity, increased bandwidth availability, and the growth of video delivery through multiple media are all driving these industry shifts, as stated by Pavlik [5].

Remarkable from a European (law) perspective is the fact that the 1996 Telecommunications Act contains many content provisions—for example, the provisions dealing with decent and indecent behavior and the provisions that enable the use of the *V-chip* in video equipment. From a European perspective, a bill on telecommunications should regulate the provision of telecommunications infrastructure and services and not so much the (types of) messages provided for by the infrastructure or service. See Chapters 2 and 3 on the division between the different "layers" of using ICT (information and communications technology), its technologies, industries, services, and markets. This chapter will not deal with content-like provisions as they are incorporated in the 1996 Telecommunications Act, such as decency provisions.

The new telecommunications act is lengthy and contains many regulatory details. As in other adult democracies, a bill will only gain sufficient support from all sectors, industries, and parties involved when there are enough provisions to satisfy all interests. It is, therefore, understandable that, for example, the broadcasting industry was won over by more relaxed licensing and media concentration requirements, the ability to reserve free spectrum, greater flexibility in the use of spectrum, longer license terms, and greater likelihood of license renewals. The

---

3. Companies such as Americast, the interactive television company headed by Steven Weiswaser and comprised of Ameritech, SBC Communications, BellSouth, GTE, and the Walt Disney Company.

telecommunications industry, on the other hand, was won over by the relaxation of solely providing typical telecommunications (telephony) services either on the state (local) or federal level.

Before going into more detail on these subjects, we will discuss the largest renewal in the telecommunications act, which is found in Title IV, Part V and is entitled "Video Services Provided by Telephone Companies." It is these provisions that really move the new regulatory environment toward a convergence of technologies and industries, bringing a real multimedia environment to the foreground and repelling the statutory restrictions against the *local exchange companies'* (LECs') provision of video programming within their telephone service areas, as well as the FCC's video dial-tone regulations. Part V supplies a new regulatory regime for common carriers that compete in the video market, calling them *open video systems.*

The new regime contains three main provisions, listed as follows:

1. Open video systems or cable systems operated by common carriers are not subject to Title II (common carrier) nondiscriminatory access obligations;

2. A detailed prohibition against cross-ownership applies to both LECs and cable television systems, generally forbidding either a LEC or a cable television operator with overlapping service areas from owning 10% or more of the other. Joint ventures between the LEC and the local cable operator to provide video programming or any telecommunications services are also prohibited. There are several exceptions to the cross-ownership prohibition for rural telephone companies, small cable systems, and cable systems subject to competition. An exception may also be granted by FCC waiver;

3. Common carriers operating or establishing a system for the delivery of video programming are not required to receive prior FCC authorization.

The broadcasting industry was allowed more *spectrum flexibility*. This gave television broadcasters an opportunity at free spectrum for a possible new generation of digital broadcast technologies and permitted them to use their spectrum for other services besides television broadcasting.

However, there was some controversy over this issue, centering on the giving out—rather than selling—of broadcast spectrum airwaves. Under the provision, holders of existing television broadcast licenses would receive 6 MHz of digital bands on loan, enabling them to run their signals on both digital and analog bands while the market for the new digital system grew. This was usually referred to as simulcasting. Broadcasters would be able to keep the digital bands but within 15 years would have to return the analog bands, which would then be auctioned off by the government. This would mean no federal revenue from the sale for 15 years to come.

While the additional 6 MHz of spectrum is the equivalent of one new channel, new digital transmission technology could now subdivide 6 MHz into multiple broadcasts or other uses, such as cellular telephones or personal communications services. The "broadcast flexibility" provision allowed television stations to use any new broadcast frequencies they received for services beyond the transmission of a single channel, by applying these new technologies.[4] Competition in this area might come from somewhat unexpected quarters such as the wireless (telephony) operators. It is now fairly certain that new wireless technologies available at the start of the next millennium will be able to deliver wideband DTH services. Against this background, the acquisition of *personal communication system* (PCS) licenses in 1996 on the spectrum auctions held by the FCC offered a glimpse of the competition that might appear between the traditional wireline operators, the CATV operators, satellite (both DBS and *low Earth orbit* [LEO]) and wireless (PCS) operators. Within three to five years, PCSs will be capable of delivering wireless video to mobile or stationary television/computer receivers. The spectrum auction was intended to open doors to entrepreneurial "small businesses." However, it turned out that just five companies bid more than $8 billion for spectrum covering more than two-thirds of the U.S. population.[5]

---

4. Regarding schemes for assigning transition channels to eligible parties [6].

5. Four of the companies are backed by Korean and Japanese investors. The largest single bid came from eight-month-old Nextwave Communications, Inc., which bid more than $4 billion for a service area covering over 40 percent of the U.S. population. Nextwave is backed by Japan's Sony Corporation, four Korean companies, and the South Korean government [7].

4.2.2 Direct broadcast satellite

DBS may also in the near future prove itself as a means of distributing multimedia programming. DBS is also exempted from certain sections of the bill, notably the ones on rates. In addition, DBS providers are exempt from any locally imposed taxes or fees; this exemption reflects the technical design of DBS as being a national or regional service, not a local service. In reality, however, new DBS technologies may well be able to start delivering local services. It is fairly obvious that the ability to provide a real local service would fundamentally transform DBS service. The inability of DBS to deliver local programming has been a major barrier to direct, all-out competition between cable and DBS.

As of September 1996, the nation's three leading DBS providers claimed a total of 3.4 million subscribers: DirecTV Inc./U.S. Satellite Broadcasting Co.—which is owned by General Motors Corp.'s Hughes Electronics Corp. and offers its subscribers 200 channels of entertainment and informational programming [8]—claimed 1.8 million subscribers. Primestar Partners, a medium-powered DBS provider owned by TCI, Time Warner, Cox, Comcast, Continental, and General Electric that offers up to 95 channels to its subscribers [9] boasted 1.43 million subscribers. Finally, EchoStar Communications Corp., which offers 40 U.S. cable channels, 30 audio channels, and up to 13 premium channels, claimed 160,000 subscribers. With the launch of a second satellite, EchoStar expects to boost capacity to 160 channels. As of September 1996, EchoStar was actively seeking a strategic alliance with other companies and was engaged in discussions with Sprint, Jones Intercable, and Lockheed Martin Corp. [10].

AlphaStar Television Network initiated DBS service (1996) in the United States while a MCI/News Corp. DBS venture is slated for 1998. Some analysts predict that DBS subscribers could top five million by the end of 1996, and estimates range from 10 to 21 million subscribers by 2000 [11].

4.2.3 Media concentration and foreign ownership

Media concentration and foreign ownership is dealt with in Title II of the new bill. Television networks, now limited to owning stations that reach a total of 25 percent of the nation's televisions, are allowed to reach

35 percent. A cap on the number of radio stations that could be owned nationwide has been lifted, although there are some limits on how many stations a single company can own in each market. The FCC's prohibition on television networks owning a cable television company (or vice versa) has also been lifted. However, the FCC's rule prohibiting same-market TV station/cable system cross-ownership has not changed. Over the objection of some key House Republicans, negotiators deleted at the eleventh hour a provision that would have allowed foreign companies to own American broadcast stations. The three-year cable antitrafficking restriction, which prevented the sale or transfer of a cable system within three years after a prior sale or transfer, has been eliminated.

The renewal policies and terms for broadcast licenses can also be found in the new bill. These provisions extend the term for broadcast licenses from five to eight years for television stations and from seven to eight years for radio stations. In addition, incumbent broadcasters receive the right to apply for renewal without competing applications, as well as a presumptive right to renewal if they have served "the public interest, convenience, and necessity" and have not committed serious violations of laws or rules. Only after denying a renewal application can the FCC accept and consider competitive applications for the license.

DTV is dealt with as follows. To be eligible for *advanced television* (ATV) licenses, a company initially has to have an existing TV license or permit. The act requires that either the original or the new license be surrendered but leaves the FCC to determine when and how. Ancillary services will have to be allowed by the FCC for ATV licensees, but such services cannot degrade ATV service and are not entitled to must-carry rights on cable. All ancillary services must meet public interest standards. Violations of FCC rules stemming from ancillary services are to be considered at renewal. The act requires the FCC to assess a spectrum fee for any ancillary service that provides compensation to the licensee (other than from advertising on nonsubscription services). The fee is supposed to reflect the value of the spectrum used for the remunerative service but is not to exceed the amount that would have been realized from an auction of that spectrum. Every two years, during a period of 10 years (up to 2006), the FCC is required to evaluate the ATV service and to determine whether the abolishment of the assigned analog transmission of NTSC broadcasts can be ended. It is directed specifically to assess consumer willingness to purchase ATV receivers, alternative uses of ATV frequencies (including public safety services), and efforts to reduce the amount of spectrum assigned to licensees. Every broadcaster will, at that time, have to turn in one of its frequencies, the channel for DTV, or the channel for analog National Television System Committee (NTSC) transmissions. The Balanced Budget Act of 1997 opened the possibility of prolonging the transition period if a number of conditions are met. Among these conditions is the failure of one or more of the largest television stations to begin broadcasting DTV signals due to causes outside the control of the broadcasters. The transition period will also be prolonged if less than 85% of the television households is able to receive DTV signals off the air either with a DTV set or with an analog set equipped with a converter box or by subscribing to a cable-type service that carries the DTV stations to the market.

From these last provisions, it can be asserted that the exchangeability of *broadcast* spectrum and *telecommunications* spectrum is becoming a reality. This assessment can also be derived from the fact that the FCC, on numerous occasions and in different documents, proposes that DTV will free parts of the broadcast spectrum for public safety as well as valuable business uses. The FCC is accelerating the successful introduction of DTV in the United States by rulemaking, enabling most Americans to gain DTV access by 1999, with a scheduled full U.S. coverage by 2002.

4.2.4 Cable television services

CATV services in the United States are seen as the new motor behind the electronic highway developments. CATV operators are expected to initiate a renewal of the local and regional markets.[6] Under the enactment of the new bill, CATV operators are no longer excluded from providing telephone services to their clients, and the telephone companies are no longer told to keep away from the "video market." Thus, at the heart of these provisions, the elimination of the ban on telephone companies offering video services and the deregulation of cable rates can be found. The first goal is consistent with the bill's central purpose of fostering competition by removing barriers between what are now distinct industries. The easing of cable rate regulations is viewed as necessary to spur CATV

---

6. In 1996, the number of homes passed by cable grew to 96 percent of all television households in the United States, and as of 30 June 1996, 64.6 percent of all homes passed received basic cable service (roughly 62 million homes). See [12, 13].

competition with the BOCs. The CATV industry asserts that it will have to make expensive upgrades in cable systems to offer telephone services and that the price controls discourage banks and Wall Street from providing the necessary financial means.

Two revisions in Title VI, which regulates the provision of video programming in a competitive environment, attempt to provide for a level playing field between new entrants and traditional cable operators in the video delivery marketplace. The revisions entail the following:

▶ Amending certain aspects of the current regulatory structure for traditional cable operators by easing the regulatory burdens of the 1992 Cable Act, thus allowing cable operators to respond more effectively to the introduction of new competition;

▶ Establishing a regime of regulation for video dial tone and other common carriers of *open video systems*.

While the legislation modifies or even eliminates several cable television *rate provisions* of the 1992 Cable Act, it does not completely overturn the current regime of rate oversight and regulatory control. For example, the legislation abolishes the *cable programming service* (CPS) rate regulation, while keeping intact rate regulation of the basic tier. Small cable operators (defined as those serving less than 1% of all U.S. subscribers and not affiliated with an entity whose gross annual revenues exceed $250 million) in franchise areas of 50,000 or less subscribers are immediately exempt from CPS rate regulation, as well as basic tier regulation if the basic tier was the only tier regulated as of December 31, 1994. The "effective competition" definition is broadened to include cable operators subject to competition from comparable LEC (or affiliated) video programming services. Operators that demonstrate "effective competition" are exempt from CPS and basic tier rate regulation. Uniform rate requirements do not apply to systems subject to effective competition, per-channel, or PPV offerings, or bulk discounts to multiple dwelling units. Subscriber notice of service or rate changes may be provided by any reasonable, written means. No notice to subscribers is required if the service or rate change is the result of a change in a federal, state, or local fee. Cable program access requirements are applied to video programmers in which common carriers and their affiliates have an attributable interest. Commercial must-carry markets are to be based on viewing patterns. The FCC must decide within 120 days on petitions to modify a commercial broad-caster's market designation.

Finally, the legislation opens the way for cable operators to enter the local exchange market. The legislation pre-empts local laws that require the cable operator to obtain a franchise before offering telecommunications services. In addition, the legislation generally pre-empts local or state laws that are intended to impede or restrict a cable operator's provision of telecommunications services.

4.2.5 Video programming regulations

Interesting from a European point of view are the rules called *video programming regulations*, which are, in reality, provisions that ensure compatible CA obligations. The legislation requires the FCC to promulgate rules to ensure the availability of consumer video programming access equipment. FCC (American) regulations are to provide for the commercial availability of subscriber equipment (converter boxes, interactive communications equipment, and set-top boxes) that subscribers may use to access video programming services. These regulations are intended to permit subscribers to have more choices in access equipment and to prevent program distributors, such as cable operators or distributors of open video systems, from forcing subscribers to use only the distributor's access equipment. Video programming distributors are not prohibited from charging subscribers for the use of their subscriber equipment, but the charges are to be separately stated and are not to be subsidized from other services. In Europe the same type of *unbundling* obligations can be found in the *open network provision* (ONP) regulations of the European Union. The Telecommunications Act directs the FCC to abandon these regulations when the FCC determines that the video services market and the access equipment market are fully competitive. Refer to Chapter 6 for a further explaination of the ONP regulations.

4.2.6 Practice of forbearance

The FCC's *practice of forbearance*, which had a long-standing legal tradition, was ruled unlawful by the U.S. Supreme Court in 1994.[7] Under this

---

7. Abandonment of tariff filing requirement exceeds the FCC's limited authority to "modify" requirements of the Communications Act. See [14].

practice, the FCC forebears from applying a statutory provision or regulation to a carrier or a service if the FCC determines that enforcement is not necessary to protect the market or consumers. This forbearance needs to be consistent with the public interest. A principal new element in this bill is that, in unambiguous terms, it grants forbearance authority to the FCC. Under certain conditions and circumstances, the FCC is even obliged to forebear. This forbearance authority is designed to end unnecessary regulation during the shift from monopoly markets to a competitive environment. In addition, the title requires the FCC to conduct a biannual review of its regulations and eliminate any unnecessary regulations or unnecessary agency functions.

4.2.7   Future developments

Whether or not the U.S. Telecommunications Act will actually bring the benefits of competition to U.S. consumers remains to be seen. Warren J. Sirota gave an extremely grim forecast on what the new bill might bring to the consumer market:

> So here is what we, as consumers, are likely to see over the next several years:
>
> ▶ Cable TV rates will rise.
>
> ▶ Local telephone rates will go down for business customers.
>
> ▶ Local telephone rates for consumers will stay the same or rise.
>
> There will be consolidation among the large players in their traditional segments. Beyond that, the crystal ball is clouded by Congress' lack of vision and obfuscation of the important issues. The bottom line is that real competition and new innovative services will not arrive for many years. When they do, very large carriers will be the only service providers. Innovative services and price competition in the consumer segment will first occur in the markets that are upper and middle class on the socioeconomic scale. Universal service will become the lowest common denominator, the lifeline, for communications service. And by the year 2000, the Decency Act will be forgotten [15].

A less grim, but nevertheless critical view on the developments "unleashed" by the new telecommunications act is given by
John V. Pavlik:[8]

> Although not the sole catalyst for the coming transformation of
> programming, the Telecommunications Act of 1996 has unleashed the
> forces making it possible. The public interest will be served, but as a
> secondary by-product of the forces of the commercial marketplace. The
> over-arching question to be decided by the harsh realities of media life
> will be whether the ideal of competition in a deregulated environment
> is really the final answer. Already, the accelerating drive to build giant
> corporate alliances and mergers points to the need for continuing
> monitoring of the media market place. There may well be an
> intensification, rather than a lessening, of governmental efforts to
> head off [a] monopoly and its market consequences.

The 1996 act removes the legal barriers prohibiting communications
companies from providing video services. As a consequence, typical
telecommunications companies will enter the programming business.
Although they have the finances, know how to handle research, and
have a substantial market reach, they do not possess the necessary tradition, infrastructure, and culture to create quality video programs. On the
other hand, the typical television networks do not have a great tradition
in producing interactive programs and using transmission media other
than the terrestrial broadcasting infrastructure. Representatives of these
industries will therefore look to each other to build (strategic) alliances.
Both industries will mount major efforts to develop interactive programming; only under these circumstances will they be able to compete in the
new digital millennium.

---

8. Pavlik [5] is the executive director of the center for new media at the Columbia
    University Graduate School of Journalism.

4.3 The grand alliance high definition television system

The GA HDTV system is composed of the best features of all of the original four digital systems. In the November 15,1996, *HDTV Newsletter* writer Dale Cripps gave an interesting account of the different positions of the parties involved, which is still a very readable account of the standardization issues at stake at that time [16]. The GA HDTV system is a layered digital system architecture with header/descriptors designed to create outstanding interoperability among a wide variety of consumer electronics, telecommunications, and computing equipment. Interfacing with other digital systems, thus creating a real ICT environment, is one of the key advantages of the new ATV standards.

The four layers of the GA-HDTV system are the following:

1. The picture layer, which provides multiple picture formats and frame rates;

2. The compression (video and audio) layer, which uses MPEG-2 video compression and Dolby AC-3 audio compression;

3. The transport layer, a packet format based on MPEG-2 transport that provides the flexibility to deliver a wide variety of picture, sound, and data services;

4. The transmission layer, a vestigial sideband signal that delivers a net data rate of over 19 Mbps in the 6-MHz simulcast channel.

The GA HDTV system provides for multiple formats and frame rates, all of which can be decoded by any GA HDTV receiver. Progressive scan formats are provided at both video and film frame rates, in addition to an interlaced format. The system includes two main source format variations, with a different number of lines per frame. These two formats have 720 active lines and 1,080 active lines per frame respectively. For interfacing with the GA HDTV encoder, the 720-line format uses 1,280 active samples per line, while the 1,080-line format uses 1,920 active samples per line. These choices yield square pixels for all formats for interoperability with computer display systems and graphics generation systems. For interfacing with the GA HDTV prototype encoder, the 720-line format is progressively scanned with a 60-Hz (nominal) frame rate, while the 1,080-line format is interlaced 2:1 with a 60-Hz nominal field rate. The encoder input accepts 787.5 total lines per frame and 1,125 total lines per frame for the two interlaced formats.

For compression and transmission, the frame rate for the 720-line format and for the 1,080-line format can be 60 Hz, 30 Hz, or 24 Hz. To retain compatibility with NTSC, the same formats are also supported with the NTSC numbers of 59.94 Hz, 29.97 Hz, and 23.98 Hz. The 60-Hz and 59.94-Hz variations for the 1,080-line format are encoded as interlaced scanned images, while the other formats are encoded as progressively scanned (see Sections 4.2.1, 4.2.3, and 4.2.7 for a description of the discussions between the broadcasting and computer industry). In addition, the GA has agreed to develop a "migration path" to a 1,080-line, 60-frame/s (FPS) progressive scan system—one that would realize this objective as soon as technically feasible. This decision by the GA would help the group reach its ultimate goal for the GA HDTV system to be a line progressive scanning system of more than 1,000 lines, with square pixels, that would promote interoperability between the new video standard and the other imaging formats, including computers. To understand the difference between interlace and progressive scanning, see Figure 4.1.

The HDTV signals will be broadcast primarily via the unused UHF channels and the NTSC taboo channels in a simulcast mode. The spatial resolution of the picture will be at least twice that of NTSC horizontally and vertically without exhibiting the interlace artifacts and poor chrominance fidelity associated with NTSC. The introduction of HDTV would ultimately mean the luxury of free home delivery of digitally clean pictures of a quality approaching that of 35-mm movies accompanied by CD quality surround-sound. These pictures will be presented in a panoramic horizontal-to-vertical aspect ratio of 16:9 as in the movies on the new HDTV receivers.

Before the resolution of the main controversies—the most important one being the choice between progressive or interlaced scanning—FCC chairman Hundt told a Warren Publishing gathering in New York in October 1996:

Digital is needed to create a public good of free digital programs, consisting of sports, entertainment, news, free time for political

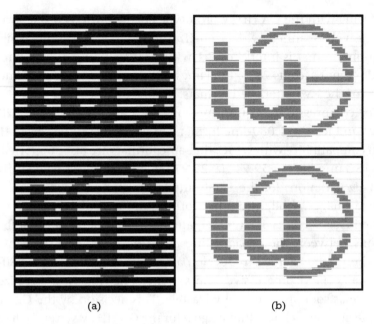

(a)   (b)

Figure 4.1   (a) Interlaced and (b) progressive scanning.

debate between presidential candidates and local candidates, educational shows for kids, public service announcements, and anything else
within reason that the public interest demands from the licensees of the
airwaves, the public's property." He concluded ominously, "If digital TV
doesn't do that, then we might as well just auction the spectrum for any
use, subject to interference taboos, and let that be our easy answer to
the tricky spectrum allocation issues posed by digital TV [16].

The success of the television industry is almost entirely due to a well-
defined single mandated NTSC (PAL/SECAM) standard. Broadcasters
and manufacturers contend that the adoption of a well-defined standard
is absolutely essential if the digital era is to be anywhere near as successful
as the analog one. Consumer Electronics Manufacturers Association said
that a standard allows the following:

▶ A national market for television receivers;

▶ Price efficiencies on receiving and production equipment for both
consumers and service providers;

▶ Plug-and-play capability as consumers move from one locale to another in the United States;

▶ "Universal service" in video delivery of entertainment, news, and emergency information across the country;

▶ Compatibility with traditional video reception and recording equipment as well as computers and computer displays.

By mid 1996 it became clear that the computer industry was late entering the DTV standardization process. One might even say that the industry hardly existed at the beginning of the ATSC process in 1987. In fact, the computer industry claimed that it was completely shut out and that an old-fashioned television model prevailed in the standard.

In 1996, it had also become clear that the computer was going to be the all-purpose communications device, and the computer industry therefore wanted to have the easiest possible interface to the most widely distributed pipeline of transmitted digital data. That pipeline was going to be the new digital broadcast standard. However, the computer industry was concerned that the ATSC standard incorporated an outmoded technology unsuitable for the future of computer-centric telecommunications. The computer industry argued that the inclusion of interlace scanning in the standard indicated a strategic disregard for the future that television would share with the computer. Broadcasters, on the other hand, argued that computer-friendly measures were included and pointed to the 720-line progressive scan and 1,080 lines at 24-, 25-, 30-MHz progressive scan formats. Computer industry antagonists held that any inclusion of interlace would perpetuate interlace displays, but broadcasters countered that the inclusion of interlace scanning is important to their live broadcast business, something the computer industry was still not capable of providing.

Broadcasters further argued that their film- and computer-generated programming is all in progressive scan format, which is also in the standard. However, they said that it was meaningless to argue the point, since any of the signal parameters are permanently decoupled from the requirements of the display. Progressive scanning can be converted readily to interlaced scanning for the display with nothing more than a low-cost chip in the receiver, and manufacturers will always need to sell low-cost receivers that use interlace display even if there is never another interlace signal transmitted in the world.

The hastily agreed standard allowed for all of the 18 interlaced scanning plus many more different formats, both progressive and interlaced [17].

4.4   Summary and conclusions

DTV is arriving. Spring 1998 saw the first terrestrial digital broadcasts in the United States. At the January 1998 Consumer Electronics Show in Las Vegas, a number of new DTV receivers were shown to the public.

At the start, these HDTV digital receivers will be expensive, approximately $5,000. However, as is common with electronics equipment, when the market really gains momentum, prices will decrease rapidly. Old NTSC receivers with an additional "set-top" box (approximately $300–$400) will be able to get some of the digital advantages on their (old analog) receivers. A much better audio and video quality will be the result and, as is usual with digital (transmission) technology, the viewer will either receive an image or no image at all. Compare this to the old analog technology where an image full of "snow" and "ghosts" would still be displayed. The real renewal is not in the displayed image, although notably in the United States it will improve spectacularly when HDTV receivers are installed and watched. The main renewal and its market effects will make the DTV a real "electronic highway" apparatus, a dominant and, eventually, a fully equipped access machine for the information superhighway. This new DTV receiver will be to the couch potatoes what the computer is to the computer whiz kids: *the* machine for access to the digital superhighway in the next millennium.

References

[1] Lim, J. S., "Digital Television: Here at Last," *Scientific American*, May, 1998, pp. 56–61.

[2] Sobel, A., "Television's Bright New Technology," *Scientific American*, May, 1998, pp. 48–56.

[3] Piper & Marbury, *Summary Of The Telecommunications Act Of 1996*, at http://www.pipermar.com/article3.html, consulted May 25, 1998.

[4] FCC, Mass Media Bureau, *Tower Siting*, http://www.fcc.gov/fcc/mmb/prd/dtv_tower_siting, consulted June 2, 1998.

[5] Pavlik, J. V., *Competition: Key to the Communications Future?*, http://www.internetgroup.com/natas/tca96.htm, consulted May 25, 1998

[6] Sixth Further Notice of Proposed Rulemaking in MM Docket No. 87-268; FCC No. 96-317, 25 July, 1996.

[7] Andres, E. L., "Asians Win F.C.C. Bidding for Licenses," *The New York Times*, May 7, 1996.

[8] McDaniel, "Dish It Out: Disillusioned by Cable, More Are Investing in Mini-Satellite Systems," *Salt Lake Tribune*, September 16, 1996, p. B1.

[9] McConville, "Primestar: First in Digital TV, First in Value," *PR Newswire*, October 4, 1996.

[10] "EchoStar, Jones, Eye Strategic Pairing," *Broadcasting & Cable*, September 30, 1996.

[11] Deagon, B., "Is Cable Industry Ready for Satellite TV Assault?," *Business Daily*, February 26, 1996.

[12] Paul Kagan Associates, Inc., *Marketing New Media*, September 16, 1996.

[13] "In the Matter of Annual Assessment of the Status of Competition in the Market for the Delivery of Video Programming," Comments of the National Cable Television Association, Inc., July 19, 1996, Appendix A, Table 1.

[14] MCI Telecommunications Corp. v. AT&T, 114 S.Ct. 2223, 1994.

[15] Sirota, W. J., *The Telecommunications Act of 1996: A Commentary on What Is Really Going on Here*, http://www.wls.lib.ny.us/watpa/telcom.html, consulted May 25, 1998.

[16] Cripps, D., "Do Not Adjust Your Set: It Is The Industry That Is Out Of Control" *HDTV Newsletter*, Advanced Television Publishing, November 15, 1996.

[17] Lims, J. S., "Digital Television, Here at Last: Avoiding a Hard Decision," *Scientific American*, May 1998, p. 59.

Japanese policy

5.1 Introduction

Developing a new television technology for deployment in society at large necessitates the involvement of a large number of participants, from government to industries and from broadcasting entities to telecommunications operators. Japan is no exception to this rule. However, there is a significant difference in the way such a research and development path is being forged within the policy, legal, economic, and technological context of Japan, compared with its evolution in other economic regions in the world such as the United States or the European Union.

Accordingly, before dealing with the history of developing HDTV in Japan, it is necessary to describe the environment in which these types of technological advances take place within the Japanese context. Section 5.2 presents a general description of Japanese policy and the Japanese legal framework used for such developments.

5.2   General policy and regulatory framework

In general, it is difficult for foreigners to understand Japanese law and policy making, including the policy making for, and legal development of, information and communications technologies. Meanwhile, the convergence of formerly different technologies, industries, and, consequently, markets affects the Japanese legal and policy-making process, mainly by creating an even more centralized policy development within the Ministry of Post and Telecommunications (MPT). Let us now examine the establishment of broadcast television in Japan and how the "system" (the parties involved) reacted to the new technological device known as the television set.

After World War II, the re-establishment of the Japanese broadcasting system occurred in 1950 under the guidance of the allied powers. Radio broadcasting was resumed by the public broadcasting system (NHK) and private-sector broadcasting stations. Three years later, NHK and *Nippon Television Network* (NTV) began broadcasting to the general public. The birth of the Japanese broadcasting system, established under the guidance of the allied powers, is exemplary of the way in which Japanese society makes policy and incorporates new (technological) developments.

Before 1985, when a more liberalized environment was created, Japan's telecommunication system was in the sole hands of the *Nippon Telegraph and Telephone Public Corporation* (NTT). However, since government regulations prevented NTT from functioning in the television broadcasting industry, television networks and NTT peacefully shared their responsibilities and refrained from interfering with each other's domain. NTT was satisfied merely collecting fees for the use of its telecommunications capacity which the television stations relied on for their nationwide broadcasting.

Newspapers and the press in general in Japan were not quick to respond to the emergence of television. At first, they failed to imagine that television would grow to be just as influential as newspapers. In addition, they looked upon television as simply a technological novelty. In the second half of the 1950s, this attitude changed, and newspapers started to consider the relationship between themselves and television. At almost the same time, the MPT devised a plan to affiliate private broadcasting stations with Japan's five major newspaper firms, and the latter agreed to carry out the plan.[1]

In Japan, television technology was—from the very outset—seen as an extension of the radio. Television broadcasting and programming was transplanted from Western society and diffused under strict government control. Moreover, HDTV, popularly known as *Hi-Vision,* was developed solely by experts without any evaluation by ordinary people and without input from amateur engineers. As is normal in Japanese technological development, a number of leading Japanese technology companies, under the guidance of the Japanese government, joined financial and expert forces for developing a top-of-the-line television technology. Again, this was done without taking "outside" developments into serious consideration (see Section 5.3 for more on this subject). The result was that Japan's HDTV technology failed to ride the wave of digitalization that was sweeping the world.

In general, Japanese law—in the eyes of a continental European lawyer—is open to many possible interpretations, and this makes it difficult to apply the law from a Western viewpoint. In a more general context, Yoshimura and Anderson depicted this behavior as follows:

> Western intellectual traditions predispose those raised within them to look for the center of a culture, a paradigm that characterizes the culture's world view. Japanese culture, however, has no system of logical principles that creates the Japanese outlook on life. Japanese culture is more like a network. It has no center, and outcome stems from the interaction of a loose web of elements. Consequently, one can't reduce the "rules of Japanese behavior" to a coherent, compact analytical framework. There isn't simply a cookbook or a flowchart one can use to understand the salaryman's actions as the product of a series of "if-then" rules [1].

Van Wolferen concluded of the Japanese legal system that the general attitude, which stems directly from a more authoritarian government, is that contemporary institutional rulings do not encourage the (ordinary) Japanese to seek justice through the legal system. In his words,

---

1. These newspapers were *Asahi, Mainichi, Yomiuri, Nikkei,* and *Sankei.*

...after detailed analysis of the contemporary legal practice one has to conclude that the law is being placed outside the "system" on purpose [2].

Thus, it seems that there is no real use in trying to discern the role Japanese law plays in the making of new technologies. It is more interesting to determine how Japanese policy making is influenced by technological progress. Again, it is useful to refer to Yoshimura and Anderson, who contend that it is cooperation that serves to curb "matching" competition. Cooperation and competition is predictably intertwined in Japanese business. According to Yoshimura and Anderson, four key ideas define this concept:

1. Competition centers on preserving market share, not on seeking profitability, and on avoiding loss rather than seeking gain;

2. Rivalry is driven by yokonarabi, the pressure to match competitors move for move;

3. Cooperation arises because fierce rivalry must be curbed by third parties;

4. Forbearance, not mutual aid, is the basis for Japanese cooperation [3].

It is in the last two points that the role of a third party, either a cartel or a Japanese ministry (usually the *Ministry of International Trade and Industry* (MITI)), comes in. In Japanese business competition, status quo is the name of the game. The Japanese companies' intention is not to win the market share from a competitor but to prevent another company from losing market share. Such a system is referred to as matching competition. "The only way to excel is to do the same things as others, in a slightly better way"[3 at 118]. In cases in which there is excessive competition, all competitors involved will wait for a third party to step in and try to work out a form of cooperation that will be acceptable to all parties. Yoshimura and Anderson point out that the model in which parties cooperate and accept arbitrage strongly depends on the context. Even bitter rivals will work together to (re-)establish "peace" in the market or between competing firms:

A company's relationship to other firms depends on the situation, so treating yesterday's rivals as tomorrow's partners is perfectly acceptable. The simultaneous prevalence of brutal competition and extensive cooperation in Japanese businesses does not reflect any uniquely Japanese emphasis on teamwork and getting along. The Japanese emphasize conformity and have developed a model for dealing with the predictable consequences of conformity [4].

It is probable that this is one of the major reasons why Japanese ministries play such a dominant role not only in (informally) settling market disputes but also in directing technology-driven innovations.

In Western organizations, effective managers usually strive to attain specific positions in the organizational structure that lets them get work done through the structure. In contrast, the way to get things done in a *kaisha* is to establish an appropriate process and atmosphere for gaining cooperation [4 at 180].

5.2.1 Regulatory environment[2]

The telecommunications and broadcasting industry in Japan is regulated by a number of laws: Telecommunications is governed by the Telecommunications Business Law (Law no. 86 of 1984, as amended); NTT, the domestic telecommunications company, and *Kokusai Denshin Denwa* (KDD), the international telecommunications carrier, are regulated by a special law, the NTT Law (Law no. 85 of 1984, as amended); the Broadcast Law (Law no. 132 of 1950, as amended) regulates the broadcasting industry; and, of course, the Cable Television Broadcast Law (Law no. 114 of 1992, as amended) and the Radio Law (Law no. 131 of 1950, as amended), also play important roles. The MPT is the regulating ministry. In addition to these laws, the so-called administrative guidance procedure plays a very important role in the aforementioned industries.

The concept of administrative guidance is somewhat strange to either American or European trained lawyers. Within the latter jurisdictions, it is common practice to consult governmental bureaucrats to interpret certain rules and to ask what standpoint they would take in certain matters. In the end, nevertheless, decisions are usually left to courts of law. As

---

2. Most data and information in this paragraph is sourced from [5].

Geist points out, "[I]n Japan, however, the role played by bureaucrats, particularly at the Ministry of Finance and MITI, cannot be overstated, since laws in Japan are generally broadly drafted with their specific intent left open to interpretation"[5]. Similarly, another writer says, "...informal enforcement is not *a* process of governing, but has become *the* process of governing. It is used to implement nearly all bureaucratic policy, whether or not expressed in statute or regulation, at all levels of government and all administrative offices" [6].

As Geist indicates, administrative guidance is presented in three different forms:

1. Nonbinding recommendations from ministry officials that are authorized by statute;

2. Informal guidance from ministry officials where statutory authority provides for formal action to achieve similar regulatory goals;

3. Informal guidance based on Ministry's subject-matter competence.

In addition, administrative guidance can be issued at two different levels:

1. The individual level, where a specific party may be asked to act in a certain manner, such as refraining from executing a transaction;

2. The industry level, where all firms within an industry are asked to act in a certain manner, such as the limiting of production of a certain product [5].

The reform of 1985 established a new regime for the telecommunication industry in Japan. The *Telecommunications Business Law* (TBL), which is administered by the MPT, divides telecommunications services into *type I carriers*, a prohibited industry, and *type II carriers*, a discriminatory industry. Type I carriers include those carriers providing telecommunications services through a telecommunications network, such as the installation of transmission lines, satellites, fiber optics, microwave, or *value-added networks* (VANs). Type I carriers are required to obtain the MPT's permission to operate, and such permission is not granted to firms whose foreign ownership exceeds 33 percent. Type II carriers can be further divided into two categories—general and specific carriers. *General carriers* are defined as all telecommunications businesses not otherwise covered by the type I or type II special carrier regulations. General carriers usually provide telecommunications services to individual companies or groups of companies and require only a notification to the MPT describing the types of services to be provided. The regulations covering special carriers are somewhat more complex. *Special carriers* are defined as those firms whose telecommunications services exceed a certain capacity or those that provide international telecommunications services. Under the TBL, special carriers are required to register with the MPT and provide the ministry with basic information such as a business plan and an outline of the services to be provided. The MPT is entitled to refuse registration should the applicant have previously violated TBL provisions or if the applicant does "not have an adequate financial basis and technical capability to properly perform the telecommunications business."[3]

Three main aspects changed the Japanese regime in 1985. The first change was that from 1985 onward MPT seized power and became the most important player in the sector. MPT became *the* regulator; in effect, this meant the following:

> ▶ Control over the budget and personnel shifted from the Diet (i.e., Japanese Parliament) to NTT with MPT supervision;
>
> ▶ Responsibility for price and service regulation shifted from the Diet to the MPT;
>
> ▶ Responsibility for technical regulation shifted from NTT to the MPT.

Second, MPT has achieved in gaining more control over telecommunications R&D and, therefore, became the responsible agency (ministry) for industrial policy in this industry (e.g., MPT announced in June 1995 plans to start financing infrastructure by making cheap loans available to telecommunications and cable television operators).

Third, MPT adopted a more active approach toward managing competition in telecommunications services. This meant that MPT would conduct lengthy negotiations and demand enormous amounts of data before it would allow NTT to start new services or implement new (lower) tariffs. The ministry generally favored new companies over NTT. This, for

---

3. For further details on the development and provisions of the telecommunications law, see [7, 8].

Japan's somewhat unusual active approach violated in itself the spirit of "deregulation," but it did not slow new competitive entrants. As of 1997, 127 type I and 3326 type II carriers (value-added network (VAN) service providers) had entered the market (see Figure 5.1). The new competitors had about one percent of the local telephone market, 40 percent of the long-distance trade, 35 percent of the international business, and 50 percent of mobile telephony.

MPT has guarded its overall control and leadership of the telecommunications sector (i.e., at the end of 1993, it shifted from restricting the cable television market to promoting it, announcing that it would allow cable companies to cover more than one geographical zone and to offer telecommunications services including telephony). In April 1994, MPT unveiled a plan to encourage cable operators to interconnect with other cable operators so that their area of coverage extended [9]. The Telecommunications Council reported on "Basic Rules for Interconnection" to the MPT minister in December 1996; as a result, a bill of amendments to the TBL was passed by the Diet in June 1997 and enacted in November 1997. Under that bill, type I carriers that own designated facilities (fixed local loop facilities in excess of 50% of the total number of subscriber lines at prefecture levels and intraprefecture telecommunications facilities installed as one system with those local loop facilities) must meet the following obligations:

▶ Introduction of an interconnection tariff system including interconnection charges and technical requirements;

▶ Preparation and disclosure of accounting reports concerning interconnection;

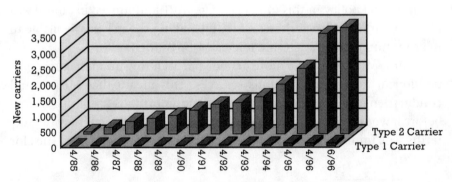

Figure 5.1   New entrants to the telecommunications industry.

▶ Preparation and disclosure of plans to revise or expand facility features or functions.

These changes to TBL were the first results of a more general policy approach toward the challenges posed by the coming of the information society. MPT stopped dealing exclusively with the telecommunications and broadcasting sector in 1994, because the 1985 liberalization did not have the intended result, and MPT's policy and regulatory scope were too limited. In May 1994, MPT through its Telecommunications Council issued a study called: "Reforms Toward the Intellectually Creative Society of the 21st century, Program for the Establishment of High-Performance Info-Communications Infrastructure." The ministry itself followed suit in January 1996 with the "Deregulation Package of Telecommunications and Broadcasting for the Second Reform of the Info-Communications System in Japan," revised in March 1996.

It became clear from the results of various studies and deregulation packages that Japan had to deal not only with telecommunications and broadcasting regulations but also with those in the area of education, health care, commercial transactions, and governmental activities. Accordingly, in January 1996 the Japanese government installed the Working Group on the Review of the Regulatory System under the Advanced Information and Telecommunications Society Promotion Headquarters. This working group will issue studies for guidance in (trans)forming the necessary regulatory environment, so that the Japanese society at large can benefit from all the revolutionary technological changes made possible by digitalization.

The proposed deregulation package covers a number of important issues, including the status and market structure for NTT and *Kokusai Denshin Denwa (*KDD). KDD was allowed to enter the domestic Japanese market. NTT, meanwhile, is to become a holding company broken up into three separate companies: one long-distance carrier (an international company), and two regional companies. The deregulation package also creates a competitive environment for new businesses and new services by rapidly implementing interconnection rules for connecting private leased lines to public ones, publishing the *Manual for Market Entry Into Japanese Telecommunications Business,* and introducing a prior-notification system and a system of standard tariff schemes for pay television and multichannel satellite broadcasting. In addition, the deregulation package establishes conditions for fair and effective competition; promotes the efficient and effective use of the radio frequency spectrum (i.e., shared use of frequency bands for both telecommunications as well as broadcasting services), simplifying licensing procedures for terrestrial multiplex data broadcasting and teletext broadcasting services; informs society of the security needs of special networks where disaster resistance is available; and assists consumers and handicapped persons, clarifying the system through which consumers can issue complaints regarding tariffs and services [10].

In a recent article, Vogel stated that although considerable progress has been made in the liberalization of Japan's telecommunications and broadcasting industry, the country still has to go a long way in really creating open competition supported by a regulatory environment. In his words:

1. Finalize the rules for interconnection between competing carriers and NTT as quickly as possible, and then move toward a forward-looking costing methodology. The MPT has already worked out a basic agreement on interconnection, and plans to work out details by the end of this year. However, the agreement still relies on NTT's historic costs to calculate interconnection charges. Under this agreement, interconnection rates will be very high by international standards. The U.S. government and NTT's competitors argue that a forward-looking approach to cost accounting (i.e., estimating the cost of building the NTT network at current costs rather than historic costs) would promote competition more effectively.

2. Require NTT to provide interconnection for the full range of services now standard in Japan—including toll-free dialing, directory services, etc., as well as basic telephone service—without charging a fee for network modification to offer the "additional" services. Competitors would be at a considerable disadvantage if they did not offer these services, and NTT should not be allowed to charge extra interconnection fees for them.

3. Require NTT to publicize technical specifications for all network interfaces (switching, signaling, transmission, etc.) so that competitors can interconnect effectively. In the future, the MPT could help prevent disputes over technical issues relating to interconnection by increasing its oversight of the process of developing network interfaces.

4. Simplify MPT licensing procedures for type I service providers. Despite incremental improvements, these procedures remain costly and time-consuming. Although the formal approval requirements are quite simple, in practice the MPT still demands detailed business and investment plans before issuing a license.

5. Ease regulations on cable television and DTH satellite services. For example, the MPT currently requires cable companies to obtain separate licenses for each franchise area, and MITI requires cable companies to power their systems for telephone signals at 60 volts (although cable companies commonly use 90-volt equipment in other countries). Likewise, in the satellite market, the MPT restricts each satellite broadcaster to 12 channels, regulates transponder lease tariff rates, and limits foreign equity participation.

6. Lift the foreign equity restriction in NTT and KDD. As part of its offer in World Trade Organization (WTO) talks on telecommunications liberalization, Japan agreed to lift the foreign equity restriction on all telecommunications service providers except NTT and KDD. But it still limits foreign equity participation in NTT and KDD to 20 percent, and prohibits non-Japanese executives from sitting on the NTT or KDD boards. Although U.S. firms probably would not seek a large stake in NTT or KDD at this point anyway, these restrictions should be removed as a matter of principle [9].

On April 1, 1998, MPT released a three-year program for the promotion of deregulation. Summarizing the proposed actions in the area of information and telecommunication business, a number of actions were announced:

▸ *Regulations for the type I telecommunications business:* The requirement for approval of each end user charge change will be abolished and replaced in principle by a notification system. A price-cap regulation will be applied to end user charges for basic services including subscribed telephone service in the regional telecommunications market.

▸ *Regulations on interconnection of networks:* The opinions of the interested parties on the introduction of *long-run incremental cost* (LRIC) methodology will be coordinated to decide the handling of LRIC by the end of fiscal 1999, based on the results of the interconnection accounts of fiscal 1998, and other measures will also be taken, thereby promoting reduction of interconnection rates.

▶ *The status of NTT:* The reorganization of NTT will steadily be implemented. While the focus is placed on the progress of implementation, effective measures will be worked out to realize the substantial competition resulting from the break-up of the East regional company and West regional company, depending on necessity.

▶ *Regulations on KDD:* The KDD Law will be abolished, and KDD will undergo complete privatization.

Also, in the broadcasting area, a number of (policy) actions have been announced:

▶ *Utilization of statistical multiplexing:* Statistical multiplexing among channels operated by program supplying broadcasters using the same satellite transponder will be utilized within the calendar year of 1998, with due consideration given to ensure fairness among program broadcasters and other factors.

▶ *Foreign investment for cable television:* As a result of the WTO agreement on telecommunications services all restrictions regarding prohibition and non-Japanese employees was abolished in February, 1998. The foreign investment limitation (not more than 25%) however, still exists for KDD and NTT.

▶ *Simplification of the permission procedures of cable television:* By eliminating details other than those pertaining to the important matters related to the business and infrastructure development plans of applications for permission to build cable television broadcasting facilities, the permission procedures will be streamlined.

The ministry's policy is directed by its approach toward the broadcasting and telecommunications business; in its view the convergence of the industries can no longer be avoided. The ministry, therefore, proposes a number of actions, listed as follows:

▶ *Utilization of the telecommunication carriers' optical fiber networks for cable television:* On the premise of ensuring fair and effective competition, optical fiber networks of subscriber lines of telecommunications carriers will be utilized as CATV transmission lines.

▶ *Radio station licensing and inspection:* A study will be conducted on the utilization of intelligent traffic systems technology on the highway and on other systems, such as vehicle operations management, to determine whether the intended frequencies can be used without causing interference.

This three-year deregulation program again means that MPT has continued adopting policies for a more competitive market in Japan, and that foreign entrance should become easier. Meanwhile, however, reciprocity in international trade is precluding a mature and competitive market. The Japanese market for telecommunications and broadcasting services is still a pretty closed market, and it remains to be seen whether some of the already implemented regulatory changes and announced plans of MPT will really create an open market in which competition is the major challenge.

Table 5.1 contains a timeline regarding the Japanese telecommunications regulation reforms of the last decades.[4]

The first results of lifting the restraints in the Japanese market can be assessed from the announcement of an increase in cooperation between different players in the area of telecommunications. For example, overseas telecommunications operator KDD and the Japanese long-distance operator Teleway Japan, a subsidiary of the Japanese car manufacturer Toyota, have agreed to merge by October 1, 1998. The merger would follow the full privatization of KDD, which the MPT had carried out in the summer of 1998. It is envisaged that the 20% ceiling on foreign ownership of KDD would be lifted on this occasion. The alliance, which would mark a further consolidation of the Japanese telecommunications industry triggered by the introduction of competition in the sector early in 1998, would follow a recent $500 million U.S. merger between the private domestic and overseas operators *Japan Telecom* (JTC) and *International Telecom Japan* (ITJ). The new KDD merger will become fully operational by January 1, 2000.

The trend toward consolidation aims at reaching a critical mass to face competition by the leading Japanese domestic operator NTT. The new

---

4. Updated but also adapted from an appendix of [9].

Table 5.1
Reform Timeline

Date	Event
June 1971	A group of young MPT officials calls for telecommunications liberalization and the reorganization of NTT
December 1980	The United States and Japan sign the first NTT Procurement Agreement, whereby NTT procurement should be open, transparent and competitive
July 1982	The Provisional Council on Administrative Reform recommends liberalizing the telecommunications service market and privatizing and breaking up NTT
October 1982	The government allows small businesses to set up VANs
October 1984	The government establishes a quasi-independent body for testing telecommunications equipment
December 1984	The Diet passes major telecommunications reform legislation, introducing competition in telecommunications services, granting MPT broad regulatory powers, and privatizing NTT
April 1985	The reform laws go into effect
June 1985	The MPT grants the first five type 1 licenses to three long-distance carriers and two satellite carriers
July 1986	ITJ becomes the first competitor in international telephone service
October 1986	The government sells the first block of NTT shares
February 1987	The MPT grants mobile telephone licenses to two consortia
June 1989	The United States and Japan sign an agreement whereby Japan will ease regulations and increase opportunities in paging and cellular service markets
March 1990	The government completes its first five-year review of the reforms, postponing any decision on breaking up NTT for another five years
March 1994	The United States and Japan reach an agreement on cellular telephones whereby *Nippon Idou Tsushin Corporation* (IDO), a service provider using a North American-type analog system, will expand its facilities to meet growing demand
January 1996	The MPT announces a deregulation plan that includes setting up a new framework for interconnection with NTT and easing foreign equity restrictions
March 1996	The ruling coalition decides to postpone a decision on breaking up NTT, but agrees to continue deliberations with the aim of reaching a decision by the first Diet session of 1997
December 1996	The government announces a decision to break up NTT into three companies—one long-distance carrier and two regional local carriers—within a holding company structure
December 1996	The Telecommunications Council reported on "Basic Rules for Interconnection" to the MPT minister, and, as a result, a bill of amendments to the TBL was passed by the Diet in June 1997 and enacted in November 1997

Date	Event
March 1997	NTT is to become a holding company, broken up into three separate companies—one long-distance carrier(an international company), and two regional companies
April 1998	MPT announces a three-year program for the promotion of deregulation regarding matters related to the information and telecommunications business

Japanese telecommunication landscape is now set to be dominated by four players: NTT, KDD-Teleway, JTC, and private long-distance and mobile operator DDI. A further consolidation of the industry is not unlikely, particularly in consideration of the fact that NTT so far seems to be the primary beneficiary of liberalization. Indeed, it is the only operator that posted a significant profit increase in 1997.

5.2.2 Conditional access and digital broadcasting

This subsection will give special attention to the issue of CA and the approach that Japan has taken so far. It is in examining CA alone where one can witness the convergence of the different industries in practice. The launch of PerfecTV[5] in June 1996 made MPT ask existing and expected digital DBS service operators to study the possibility of implementing a universal integrated receiver and decoder. At first, JSkyB and PerfecTV decided to give access to each other's decoders; later, they decided to merge their broadcasting activities as of May 1, 1998.[6] The move leaves only two rival DBS systems, the second being DTVJ, which is led by America's Hughes communications. JSkyB is owned by the U.S. media giant News Corp. and the Japanese software, television, and electronics groups Softbank, Fuji TV Network, and Sony. The Japanese trading houses Mitsui, Itochu, Sumitomo, and Nissho Iwai lead PerfecTV. Further competitors would include Toei Channel, a DBS service that

---

5. Owned by Japan Satellite Systems and the trading companies Mitsui, Itochu, Sumitomo, and Nissho Iwai, PerfecTV! offers 99 television and 106 audio channels.

6. The five main shareholders with an 11.4-percent stake each are the U.S. media giant News Corp. and Japan's Fuji TV Network, Itochu, Softbank, and Sony.

went on the air in July 1998, and a new service to be launched in 2000 by Japan's national broadcaster NHK.

MPT expects that other digital service operators will make their service offerings as compatible as possible. DirecTV, for example, is planning to develop a decoder compatible with the PerfecTV and JSkyB ones.

The privacy issue also concerns the Japanese government. Accordingly, in September 1996, MPT established guidelines on the subscriber's personal information in broadcasting for all broadcasting services. The latter ones include, of course, the digital broadcasting services as well as those service providers that provide CA services. The Japanese, in implementing these guidelines, followed the 1980 OECD privacy protection guidelines and therefore set a minimum standard for handling subscribers' personal information.

MPT is also preparing plans to introduce terrestrial digital broadcasting in Japan starting in the Tokyo area in the year 2000 and progressively covering the whole country by 2006. Licenses would be issued to either program producers or broadcasters.

Meanwhile, a group of Japanese companies have unveiled plans to establish a joint venture in 1998 to introduce DBS services via portable receivers, mainly for car use, by 2000. The main backers of the venture, which would offer about 40 digital radio and television channels, are Toshiba, Kenwood, Mitsui, JSAT, Tokyo FM Broadcasting, and Nippon Broadcasting System.

Yet another consortium has been formed in Japan to develop new technologies in order to give consumers wideband access capabilities in their premises equipment. To become leaders in this technology, a group of 20 leading high-technology companies have announced that they intend to test in Japan in 1998 a technology that allows for the high-speed transmission of data over wireless *personal handyphone system* (PHS) circuits. While PHS currently offers a 32-Kbps transmission speed, the new technology would increase that to 25-30 Mbps. Involved companies would include Japanese telecommunication operators NTT, KDD, and DDI; Japanese electronics groups NEC and Fujitsu; and possibly the U.S. and European telecommunication manufacturers Motorola and Ericsson. The Japanese MPT, at an informal meeting (August 1997) during the *Asian Pacific Telecommunication Community* (APT) which is comprised of 29 member countries and big corporations such as NTT and KDD, that an agreement had been reached to work toward common standards for the next generation of (wide-band) mobile phones. Work would start in 1998 with the aim of setting Asian standards as global standards recognized by the *International Telecommunication Union* (ITU) instead of European- or United States-led standards.

5.3 History of developing HDTV

It is only fair to state that the Japanese were the first to start thinking about what the next generation of television broadcasting should look like. Already in 1964, Japanese broadcaster NHK began its first research devoted to improve the "poor" quality of the NTSC television broadcast. NHK engineers were not happy with the broadcast quality of the 1964 coverage of the Tokyo Olympic Games and launched research activities on next-generation television technologies. However, NHK was not legally allowed to engage in producing television sets. Therefore, in 1970, it formed a coalition of manufacturers to engage additional research capabilities and to have a platform that would really produce the needed apparatus in case the research proved viable.

During the mid 1980s another group investigated the possibilities of enhancing image quality regarding the then available television sets. This group, *Broadcasting Technology Association* (BTA), developed the *extended-definition television* (EDTV), sometimes also referred to as *Clear Vision*. This system gave a 60% enhanced horizontal resolution and a 30–60% improved vertical resolution; furthermore, the EDTV was completely compatible with the NTSC standard.

The BTA was developing EDTV with an association of 12 broadcasters (including NHK) and 14 television makers (including Philips Japan). EDTV aimed at realizing all necessary compatibilities, which is understandable considering the commercial nature of the parties involved. In the opinion of the aforementioned parties, the development of a new television system that was incompatible with earlier generations was not a viable commercial option. EDTV did not have a real impact on HDTV developments, except for the fact that involved parties met and knew in some detail each other's directions.

As a result of all these efforts, 1984 saw the presentation and agreement of all parties involved in completing a set of new standards for the first HDTV, called *Hi-Vision*. At that time, Hi-Vision was also technologically ready to be produced. Without software (i.e., films), however, a new technology in itself is not enough. Accordingly, NHK attempted to win American broadcasting and film industry's approval of the Hi-Vision standard. In negotiations with representatives of these industries, a number of parameters that came with the Hi-Vision system (i.e., 1,125 scanning lines, 59.94 fields per second for interlaced displays and a 5:3 aspect ratio) were changed (to 60 fields per second and an aspect ratio of 16:9). In retrospect, it seems that this result from the negotiations between the industry representatives came back to haunt NHK, because the modifications were seen as an effort to make Hi-Vision completely incompatible with NTSC and PAL/SECAM standards [11].

Japan's aim regarding HDTV has always been to set a new world standard that would put an end to the incompatible analog standards (i.e., PAL, SECAM ,and NTSC). As a consequence, the Hi-Vision standard was (completely) incompatible with all these systems. From an industrial viewpoint, this Japanese approach can well be understood. It is, however, a bit naive to assume that two other major economic regions (i.e., Europe and the United States) would adopt such a standard without their own industry being involved in the technological development of the system. Nevertheless, there was a small period in time where it appeared possible that Europe would consider implementing the Hi-Vision standard. This was the result of NHK efforts to promise a cheap converter ($100) from HiVision toward PAL.

The real "proof" for the Japanese system came in the CCIR Dubrovnik meeting of May 1986 where the compromise between Japan, Canada, and the United States was presented and the parties involved were convinced that the Japanese HDTV standard would become a world ITU (CCIR) standard. To the surprise of a number of parties, not in the least the Japanese, the European countries voted against the proposal. The Japanese were disappointed and angry. They had expected a more generous "welcome" for their system, because as part of their policy they had promised free access to the NHK HDTV patents, and they had continuously informed all interested parties in the making of the HDTV standard proposal.

The disappointment over the Dubrovnik meeting was no reason for Japan to stop striving toward a strong local HDTV market. As a matter of fact, most new products have always relied on a strong Japanese home market. Accordingly, the next move of the Japanese government concerning the Hi-Vision developments was to be expected. July 15, 1988, saw the establishment of the *High-Vision Promotion Center* (HVC), a public service corporation approved by the MITI. Its goal was formulated in a leaflet of HVC:

> The Center promotes the utilization of Hi-Vision through the extraction, verification, investigation, research and analysis of problems existing in the public service including museums, medicine and education, and industrial areas including theaters and amusements. This is being implemented through close communication and cooperation between Hi-Vision users and the hardware/software manufacturers.

The MPT nominated 14 "Hi-Vision Model Cities," which received financial support for promoting Hi-Vision. Research was then conducted to determine how users would react and how popular HDTV hardware and software would become in these cities. Apart from MPT's "Cities" initiative, the MITI developed a Hi-Vision Community Concept under which regions were selected to become Hi-Vision communities and then granted financial support to promote the development of diverse Hi-Vision applications. These areas were also able to lease HDTV equipment inexpensively.

At the end of the 1980s, a number of different new (digital) television technologies and accompanying standards were being researched and/or implemented. Table 5.2 presents an overview of the situation by 1989.

5.3.1 MUSE

To enable the transmission of a Hi-Vision signal, 20 MHz was needed. Compared to the old NTSC system approximately five times more information was incorporated into the Hi-Vision signal (NTSC used 4.2 MHz). In Japan, the terrestrial frequencies assigned for television transmission was a maximum of 6 MHz. This property meant that terrestrial television broadcasting could not be realized and that satellite transmission was needed for Hi-Vision broadcasts. This again meant trouble for Japan, because the number of available channels in the allotted frequencies for DBS was only eight and the bandwidth of these DBS satellite frequencies was comparable to the terrestrial bandwidths. Compression techniques

Table 5.2
Overview of Different Television System Standards (Under Development) in 1989

Name of Standard	Country	Scanning Lines	Active Lines	Frequency	Scanning Method	Aspect Ratio	Bandwidth (MHz)	Sampling per Active Line
NHK	Japan	1,125	1,035	60	2:1	16:9	30	1,920
EU95	Europe	1,250	1,152	50	1:1	16:9	60	1,920
HD-NTSC	United States	1,050	—	59.94	1:1	16:9	6 + 6/3	—
HD-MAC60	United States	1,050	—	59.94	1:1	16:9	9.5	—
MUSE	Japan	1,125	—	60	2:1	16:9	8.1	—
HD-MAC	Europe	1,250	—	50	2:1	16:9	12	—
D-MAC	Europe	625	—	50	2:1	4:3	12	—
D2-MAC	Europe	625	—	50	2:1	4:3	9	—
PAL	W. Europe-France and E. Europe	625	—	50	2:1	4:3	5	—
SECAM	France and E. Europe	625	—	50	2:1	4:3	5	—
NTSC	United States	525	—	59.94	2:1	4:3	4.5	—

were needed to bring about a solution. In January 1984, NHK announced MUSE, which enabled the compression of an originally 20-MHz channel to approximately 8 MHz. This compression allowed Hi-Vision transmission over a single (27-MHz or 24-MHz bands) DBS channel.

MUSE is looked upon as a compression technique, but the original image is divided into four parts (together forming one image) that are transmitted consecutively, and the MUSE decoder in the receiver reconstructs the image by putting the original image together again. The MUSE technique is depicted in Figure 5.2.

5.3.2 Wide-screen market developments

In 1996, the Japanese wide-screen market reached a production of 2.8 million sets, controlling about 28% of the television set market. The total production of color television sets in the same year was about 10 million sets.

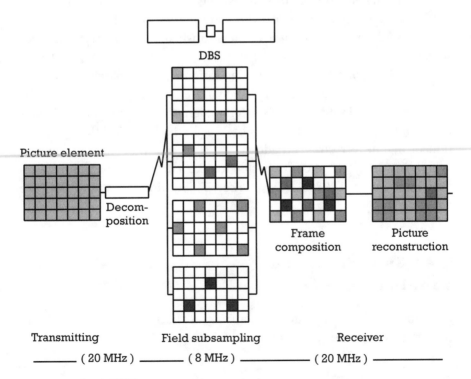

Figure 5.2 MUSE compression technique.

There are three types of wide-screen television sets: First, there is HDTV, which has 1,125 horizontal lines of resolution and is designed to receive NHK analog signals. 1996 saw the sale of about 190,000 HDTV sets, and a total of approximately 420,000 sets have been shipped since its introduction in 1991, at a set price (1997) of approximately $40,000. In the beginning, this was a very expensive television set indeed. In mid 1997, this type of wide-screen set (32-inch) could be purchased for about $3,000.

Second, there is the wide-screen referred to as *MUSE decoder NTSC-HDTV* (MN-HDTV), which provides 525 horizontal resolution lines and is capable of displaying "real" HDTV programs. The number of shipped MN-HDTV sets reached approximately 580,000 (figure from August 1997). Its set price is approximately $5,000.

Third, there is the extended-length television, which cannot receive HDTV signals but stretches the 4:3 aspect ratio into 16:9, with 525 horizontal resolution lines. In 1993, typical extended-length set prices were about $3,500; by mid-august 1997, this had fallen to $1,500. The accumulated number of shipped sets was about 8 million by mid 1997, but shipment figures of 1996 were higher than the 1997 figures. Thus, it seems that consumers are less satisfied with this type of wide-screen.

In 1997, about 30% market penetration (approximately 13 million households) for analog satellite receiving equipment was reported. A migration toward a digital HDTV satellite signal is projected for 2000. Neither real projections nor decisions have been made toward the introduction of digital terrestrial broadcasts. MPT held a first meeting on this subject in June 1997 and is expected to take a decision toward introduction by October 1998. CATV market penetration is estimated to be about 10%.[7]

5.4 Summary and conclusions

In retrospect, one can state that Japan was the first country to see the necessity for a new generation of televisions, but it was too early:

---

7. Figures and observations from [12].

The world was not ready for its system. Japanese disappointment over the results of the ITU, CCITT meeting of Dubrovnik 1986 must have been enormous, as the technological advances in the area of advanced television have come virtually to a halt since that time. Nevertheless, the long-awaited liberalization of the telecommunications and, to a lesser extent, broadcasting markets in 1998 seems to be stimulating a renewal of technological developments in this area. The most prominent example of this renewal is the no longer exclusively Japanese consortium formed to establish the wireless wide-band technologies for accessing the electronic highway—be it Internet or DTV—in the next millennium.

References

[1]   Yoshimura, N., and P. Anderson, *Inside the Kaisha; Demystifying Japanese Business Behavior*, Boston, MA: Harvard Business School Press, 1997, p.33.

[2]   van Wolferen, K., *The Enigma of Japanese Power: People and Politics in a Stateless Nation*, London: MacMillan, 1989, pp. 283–284.

[3]   Yoshimura, N., and P. Anderson, *Inside the Kaisha: Demystifying Japanese Business Behavior*, Boston, MA: Harvard Business School Press, 1997, p.105.

[4]   Yoshimura, N., and P. Anderson, *Inside the Kaisha: Demystifying Japanese Business Behavior*, Boston, MA: Harvard Business School Press, 1997, p.120.

[5]   Geist, M. A., "Foreign Direct Investment in Japan: A Guide to the Legal Framework," *9 Banking & Finance Law Review 305* , 1994.

[6]   Haley, J. O., *Administrative Guidance Versus Formal Regulation: Resolving the Paradox of Industrial Policy*, Law and Trade Issues of the Japanese Economy (eds. G. R. Saxonhouse and K. Yamamura) 1986, p.111.

[7]   Sato, H., and R. Stevenson, "Telecommunications in Japan: After Privatization and Liberalization," *24 Columbia J. of World Bus. 31*, Spring 1989.

[8]   Kosugi, T., *New Developments in the Telecommunications Industry*, in Legal Aspects of Doing Business With Japan 1985, ed. by I. Shapiro, 1985, p. 363.

[9]   Vogel, S. K., *Japanese Deregulation: What You Should Know, Telecommunications Reform in Japan*, in Japan Information Access Project, http://www.nmjc.org/jiap/dereg/papers/deregcon/vogel.html, consulted May 18, 1998.

[10] Toyoda, A., *Deregulation Policy of Telecommunications and Broadcasting in Japan*, in Japan Information Access Project, http://www.nmjc.org/jiap/jdc/cyberjapan/toyoda.html, consulted May 18, 1998.

[11] Hart, J. A., and J. C. Thomas, in: *Idate, European Policies toward HDTV: Communications Strategies, no. 20*, Montpellier, 1994, pp. 23–63.

[12] Toshitada, N., "Overview of Japanese Widescreen Deployment," at *The Digital Widescreen Television Forum*, IBC, London, November 20–21, 1997.

European Union policy

6.1 Introduction

The fact that the *European Union* (EU) has consisted of 15 culturally different member states since 1995 can only hint at the difficulties involved in developing a pan-European policy on a subject as difficult as television. These difficulties do not arise as a consequence of the fact that the television set has always been a highly technologically driven apparatus. The difficulties stem from the fact that when speaking of television, one also speaks of culture. In the EU context, this implies different cultures in as many as 15 Member States.

The European development of DTV started outside the EU competence under the so-called EUREKA flag. EUREKA was a pan-European research cooperation that resulted in, among other things, funded research projects on HDTV involving HD-MAC (*high-definition multiplexed analog component*). Later on, this technological development was supported by the EU through the incorporation

93

of the used technologies into a legally relevant text: the HD-MAC Directive. However, the obligation for European satellite operators to apply D2-MAC never attracted much support in the (audiovisual) industry and satellite operator business.

Before going into too much detail, however, it is necessary to discuss the general EU policy and the legal framework in which the development of advanced television came into being (see Section 6.2). Figure 6.1 presents the framework applied in the European advanced television's context (with reference to Section 2.3.1 on the layer model).

6.2   EU policy and regulatory environment: digital television

It is apparent that the EU is interested in a pan-European approach in the arena of converging technologies in general. In this sense, advanced television services is just one of the areas in the ICT revolution. This fact is underpinned with technological developments through the complete digitalization of industry areas that used to operate separately from each

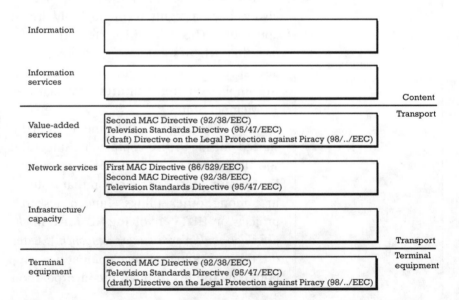

Figure 6.1   Layer model and EU policies on advanced television.

other. The 1997 Convergence Green paper [1] formulated this trend as follows:

> ... the underlying trend is the common adoption of digital technologies by the relevant sectors. Digital technologies covers a range of disciplines generally associated with the computer and telecommunications industries- digital micro electronics, software and digital transmission.
>
> ... Computer technology now plays a key role in content creation and production in both cinema and broadcasting worlds. The way in which audio visual material is produced, delivered and consumed are evolving. Content is becoming "scalable" so that it can be used in different environments and delivered on different network infrastructures.

This quote illustrates the way the European Commission views the digitalization of both content and transmission capabilities. It does not give a rationale for its policy and regulatory involvement in shaping the information society. To be able to discern this rationale we need to go back to the mid 1980s. The EU's involvement started from a more general sentiment, namely the liberalization of the telecommunications industry following the U.S., U.K., and Japanese developments in this area.

The EU became committed and wanted to become more involved in deregulating the telecommunications industry. It is beyond the scope of this book to provide a general overview of EU telecommunications policy. Accordingly, this chapter will focus on the related telecommunications and general competition laws and the issues involved with advanced television services. It has to be stated that the EU is not the only bureaucratic player in this area. The EBU and ETSI (*European Telecommunications Standards Institute*) are also important participants.

6.2.1 Background EU: telecommunications

The European Economic Community was established through the conclusion of the Rome Treaty in 1957. Originally, the EU consisted of only six members. The treaty's main goal was to establish a stronger economic cooperation between countries that waged war only a decade earlier. The ultimate goal was to establish a Europe without frontiers, thus enabling the free movement of people, goods, services, and capital. In 1994, the multilateral European economic cooperation changed its designation from the *European Economic Community* (EEC) to the EU under the enforcement of the Maastricht Treaty. Subsequently, the coalition's goal, a Europe without frontiers, did not change but the membership did. In 1998, the EU has 15 Member States.

In 1984, the EU's interest in the area of telecommunications started somewhat hesitantly with regulations concerning specifications for terminal equipment. This interest in terminal equipment fitted neatly into one uncontested area of the EU's regulatory involvement, namely the free movement of physical goods across borders. When the European Commission published its *Green paper on the Development of the Common Market for Telecommunications Services and Equipment* [2], it became clear that in the coming years a hesitant position of the EU was no longer to be expected. In summary, the goal of the European Commission can be described as follows:

> In order to establish and maintain a continuing and balanced economic growth within the EU it is necessary to have a strong European telecommunication sector to realize a good competition position of the European economy by eliminating the different national rules and monopolies in the telecommunication sector.

In retrospect, one can assert that the EU indeed managed, through a large number of regulatory measures, to free the market for telecommunications services and infrastructure provisions of most of its legal barriers. January 1, 1998, marks a significant milestone. From that date, the telecommunications market in (most) EU countries was fully liberalized.

The EU liberalized the telecommunications market by using two different legal instruments made available in the treaty:

1. Introducing open and fair competition with the privatization of the state-owned public PTTs (*Post Telegraph and Telephone*) and markets;

2. *Harmonizing* the (mostly very different) technical specifications of all the Member States.

Bringing an almost fully state-controlled telecommunications market and parties into a fully free and competitive market is something that cannot be done overnight and without a sound underpinned legal view concerning general competition law. It is therefore important to clarify the EU policy and the telecommunications market liberalization by going into some detail on (general) competition law.

6.2.2   EU competition policy: telecommunications

For decades Europe did not have a fair and open market in the telecommunications market. Instead it was a fully state-controlled business. The 1987 Green paper formulated the following policy goals:

▶ Complete liberalization of the terminal equipment market;

▶ Mutual recognition of national type approvals;

▶ Increasing liberalization of the telecommunications market;

▶ Clear separation of regulatory and operational responsibilities in the Member States;

▶ Creating conditions for open access to networks and services through the ONP program;

▶ Establishing ETSI;

▶ Complete application of EU competition rules in the telecommunications sector.

In 1988 and in 1990 the European Commission issued two obligatory directives that ordered member states to end the legal monopolies on terminal equipment and the provision of telecommunication services, respectively. However, the liberalization of services was rather limited as satellite communication, mobile services, fixed voice telephony, and infrastructure were not included within the scope of the services Directive.

The 1992 review of the measures issued between 1987 and 1992 showed no sufficient effect and therefore a furtherance of both policy and legal measures to gain more momentum in the liberalization was deemed necessary [3]. The major decision, as a consequence of the review, was that a complete liberalization of all telecommunication services, including the main earner, fixed-voice telephony, had to be realized by January 1, 1998. A derogation was accepted from a number of EU countries for this complete liberalization as from January 1, 1998 (i.e., Spain, Ireland, Greece, Portugal, and Luxembourg). Ultimately before January 1, 2002, these countries will also have to comply.

The second major decision was that the ONP concept proved to be a solid basis for future regulation, notably in areas such as universal service obligation, interconnection, and access tariffs. Third, the EU promised to formulate an explicit policy regarding mobile and personal communications systems [4] and the provision of infrastructure services [5, 6].

As stated earlier, legal instruments stemming from the EU treaty were used for opening up the telecommunications market. Seen from a regulatory point of view, the fully competitive environment for the telecommunication sector was built on two important legal principles: liberalization and harmonization. Figure 6.2 illustrates the legal principles for building a competitive telecommunications sector.

In retrospect, the EU managed to reach these goals—that is to say, necessary and relevant regulations have been implemented at the EU

Goal	Competitive environment for telecommunication services	
Realized through	**Liberalizing**	**Harmonizing**
	Directive 90/388 on competition in market for telecommunication services (often referred to as Services Directive)	Directive 90/387 on establishment of internal market for telecommunication services through implementation of open network provision (often referred to as ONP-framework Directive)
	Amendments to Services Directive: Directive 94/46 on satellite communication; Directive 95/51 on Cable TV networks; Directive 96/2 on mobile communication; Directive 96/19: implementing full competition as of Jan. 1, 1998.	Based upon this Directive: ONP-Directive leased lines (92/44); ONP-recommendation Packet Switched Data Services (PSDS) (92/382); ONP-recommendation ISDN (92/C 158/01) ONP-Directive Voice telephony (95/62); ONP-Directive interconnection (97/33)

Figure 6.2 Legal principles for building a competitive telecommunications sector.

level. Until now, however, the implementation at the national level has not always been a success. It will take at least a few years into the next millennium before a complete and transparent system of pan-European regulations will be in place, so that open market access and workable competition is really established.

A citation from a G7 Summit held in Brussels in February 1995 regarding the information society states the focus of these regulations:

> ... the regulatory framework should put the user first and meet a variety of complementary societal objectives. It must be designed to allow choice, high quality services and affordable prices. It will therefore have to be based on an environment that encourages dynamic competition, ensures the separation of operating and regulatory function as well as promotes interconnectivity and interoperability.

The challenge for the EU is probably not the (legal) liberalization of the telecommunications sector, but making and/or creating adequate (regulatory, technical, and market) conditions for dealing with convergence of the telecommunications, information technology and audio visual industries.

6.2.3 Telecommunications and general competition law: digital television

Creating an advantageous environment for new technological phenomena, such as DTV, was not considered an integral part of ICT and its regulatory developments within the EU until recently. It was seen merely as the next step in the already rich history of (television) broadcasting. The integral approach of the liberalization of the telecommunications sector made it clear that the DTV developments could not be separated from the liberalization of the telecommunications sector. Thus, it became obvious that liberalization (of the former public utility like sectors such as the telecommunications industry) could not be dealt with in separation of competition law at large.

Looking at the liberalization of the telecommunications sector from a general competition law perspective, it appears that four aspects of general competition law guided the sector's liberalization:

▶ The separation, at the national level of operational, regulatory, and controlling functions in the telecommunications sector that used to be handled by only one organization, the state-owned PTT;

▶ Separating these state-owned enterprises from the state and privatizing the telecommunications companies (in most EU countries, these enterprises have now become stock exchange companies);

▶ Between 1990 and 1998, EU public liberalization policy was aimed at creating a fully competitive environment for the telecommunications sector. This was done by gradually liberalizing all telecommunications services except fixed voice, telex, and telegraphy services (in 1998 these services too were opened to full competition);

▶ In 1990, at the start of liberalizing the sector, the complete liberalization of all (based on EU-wide applicable common technical criteria) terminal equipment proved to be an important aspect in creating the fully competitive environment for the sector in 1998.

Now, in 1998, it remains to be seen whether the policy and regulatory developments in the area of telecommunications will, in time, fully incorporate the future developments of DTV services in Europe. It also remains to be seen whether these developments will lead to a completely liberalized television services environment. It can already be discerned that public service television will remain an important cornerstone of European public (cultural) policy. Separate public funding structures will either remain in place or will be recreated to enable public policy influence in this area.

6.2.3.1   Television Without Frontiers Directive

The history of the *Television Without Frontiers* (TVWF) *Directive* starts in 1989 when the Council adopted a Directive based on Article 57 of the European Union Treaty (EUT). Article 57 relates to the freedom to deliver services throughout the EU. On October 3, 1989 the Council adopted a Directive on the promotion of European television productions and advertising and the protection of minors (OJ No. L 298, 17 October 1989, p. 23). In this directive, two provisions dealt with the proportion of European works and with independently produced European works broadcast by community TV channels. The Commission published a *Communication* on these issues on March 3, 1994 (COM(94)57), a second one

(COM(96)302) and even a third one on the overall application of the Directive on October 24, 1997 (COM(97)523). The Directive was amended through Council Directive 97/36 (OJ No. L 202, 30 July 1997, p. 60). Member States now need to implement the provisions before the end of 1998. Important in relation to the scope of the Directive are the noncompulsory quotas for transmission of community television productions, but the Directive leaves news-services (e.g., video-on-demand) out of the Directive. Advertising rules also apply to teleshopping. Member States may take measures to permit broadcasting of major events to a large public. Moreover, Member States have to draw up lists of outstanding events of general interest at a low frequency. Prior to restrictions imposed to exercise exclusive rights, Member States should consult interested parties and notify the commission. Member States will have to implement the amendments to this Directive by the end of 1998.

This Directive also spawned court cases. The first was a Belgian court that asked the *Court of Justice* (CoJ) for a preliminary ruling on the interpretation of Article 2 of the TVWF Directive (Case C-14/96). On May 29, 1997, the court ruled the following.

▶ A television broadcaster falls under the jurisdiction of the Member State of establishment;

▶ A Member State may not oppose the retransmission of programs by a television broadcaster established in another Member State, even if programs are contrary to quota obligations as laid down by Article 4 and 5 of the Directive (OJ No. C 228, 26 July 1997, p. 1).

The second court case also occured in Belgium. In this case, a Belgian court asked the CoJ for a preliminary ruling on the interpretation of the TVWF Directive to determine whether the *VT4* channel, that was broadcast from the United Kingdom, was broadcasting under Belgian or British jurisdiction (Case C-56/96). The CoJ ruled that broadcasters established in different jurisdictions fall under the jurisdiction of the country where decisions on editing and programming are taken (OJ no. C 228, 26 July 1997, p. 3).

The Swedes also asked for an interpretation of the Directive in relation to television advertisements (Joined Cases Case C-34/95, C-35/95, C-36/95). On July 9, 1997, the court ruled that the Directive permits Member States to take measures against broadcasters from other Member

States to protect consumers from misleading advertisements, provided that measures do not prevent transmission. However, since the Directive provides for the protection of minors, Member States are precluded from applying their national legislation prohibiting advertisements for children under 12 years (OJ no C 252, 16 August 1997, p. 12) to broadcasters from other Member States.

6.2.3.1   Public service television

The European Parliament took the initiative to report on the role of *public service television* (PST) within the European Union. The parliamentary Committee on Culture, Youth and Media adopted a report on the PST issue on July 2, 1996. This report stresses the importance of PST and calls on the commission to do the following:

- Propose changes to the treaty to allow development of positive PST policy;
- Exclude PST from provisions of future media concentration legislation;
- Recognize the key role of PST in a forthcoming Green paper on new audiovisual services;
- Financially support European PST, such as ARTE and EURONEWS.

The European Parliament's plenary adopted this report on September 19, 1996. To be able to prepare the Amsterdam Summit on revising the Maastricht Treaty in this respect, an expert meeting was organized on February 17–18, 1997. The experts that gathered at this meeting concluded the following:

- Public service broadcasting is important in the future media landscape;
- Cooperation between Member States is necessary;
- Decisions on the financing of PST should be taken at a national level.

The expert group advised that the Amsterdam Treaty, the treaty that succeeds the EU Maastricht Treaty, should do the following:

▶ Provide for increased legal certainty regarding public TV services and State aids;

▶ Recognize the role of public service broadcasting in Europe;

▶ Exempt public broadcasters from community competition rules.

These conclusions and recommendations were forwarded to the Council of Ministers and the European Commission. A separate protocol, which was to be added to the revised Maastricht Treaty text, was formulated. During the summit held in Amsterdam June 16–17, 1997, the European Council agreed on the final text of the PST protocol, and the protocol was added to the Amsterdam Treaty.[1] The protocol allows Member States to finance television channels in return for public service obligations. The relevant public service obligations are to be defined by each Member State itself. The European Commission has to ensure that the public funding of PST services is not contrary to the community's competition rules.

Hence, it becomes increasingly clear that even PST and the way in which Member States are allowed to implement the duties and obligations accompanying such implementation may not impede competition. This principle was even "forwarded" to the culturally sensitive PST, which is deemed important in many EU countries.

6.3 Digital video broadcasting

Before zooming in on DTV, it is necessary to describe the background of technological and market developments from analog HDTV to digital wide-screen television and to explain the accompanying EU regulatory coverage for HDTV. These developments preceded the establishment of the current European technological and regulatory framework on digital wide-screen television.

---

1. The final text of the (new) European Union Treaty was signed in Amsterdam on October 2, 1997.

6.3.1 Technological developments

The EBU started its first activities with regard to HDTV in 1981, by establishing the *Working Party V*. Together with the U.S. ATSC (*Advanced Television Systems Committee*) a first concept of a world standard was drafted. The CCIR established an IWP in 1983, with the assignment of defining a world standard for producing and transmitting HDTV. In 1985, most elements for the standard were agreed upon, and most people involved believed that the CCIR meeting of 1986 to be held in Dubrovnik would indeed produce a world standard. However, controversy arose over the usage of either a 50-Hz- or 60-Hz-based HDTV system. A number of European countries, to the surprise of the United States, kept opposing the 60-Hz CCIR draft standard, and the Dubrovnik CCIR meeting agreed to postpone a decision until the next CCIR meeting was held in 1990 in Düsseldorf.

To be able to present a working European HDTV alternative at the CCIR meeting of 1990, the European Electronics Industry signed a memorandum of understanding in March 1986 seeking funds through Eureka. Eureka was a cooperation mechanism, "invented" by French president Mittèrrand, that involved 19 different European countries through which personnel and financing was directed at technological research. Eureka projects needed to be aimed at the future and to be applicable in practice. A new (European) HDTV fitted very well in this concept. In September 1986 the Eureka Council of Ministers accepted the HDTV proposal from the European industry as Eureka project number 95, usually referred to as Eureka95. In its first phase, 1986-1990, approximately 2,000 man years and 220 million ECUs were spent. A prolongation of the Eureka95 (phase two) was approved in 1989, and by late 1992 a total of 625 million ECUs and 5,000 man years had been spent [7]. The Olympic Winter (Albertville) and Summer (Barcelona) Games, the World Expo (Sevilla), and the European Soccer Championship of 1992 proved that regular HDTV transmission could be realized. The third phase started in January 1993 and ended in June 1994. This period was used to complete all technological "loose ends." Seen from a technological point of view, Eureka95 was a success.

However, technological success alone does not constitute success in the market. Different political perspectives in a number of European countries and doubts about the potential success of the HDTV system has prevented a successful market introduction. The whole situation was made worse by the fact that within the EU no successor for the Eureka project could be agreed upon. From the beginning of 1992 until about the middle of 1993 no (political) agreement was possible at the EU level. As a result of this impasse and the (digital advanced television services) developments in the United States, in June 1993 the EU Council of Telecommunications Ministers agreed to spend about 250 million ECUs, at least half of which was to be spent on producing HDTV programs. The other half was meant to be spent in furtherance of HDTV technology. The original amount of money that was to be allocated for the technological furtherance of European HDTV was approximately 1 billion ECUs. The June 1993 decision of the EU was therefore considered to be a major setback for the research and development of the technological side of HDTV.

During this time, there was one weakness visible in the European HDTV developments, namely the usage of an analog transmission path that relied on the HD-MAC and D2-MAC standards. This was considered a weakness because the United States was aiming at the development of a completely digital system. However, it has to be realized that the HDTV system in its completeness was a digital system, since about 85% of the used and defined technology was digital, and only 15% dealt with analog transmission technology.

6.3.2 Regulatory coverage for HDTV

Terrestrial broadcasting transmission norms in the EU were different: PAL and SECAM. The next technological television transmission family needed to be a pan-European norm, a transmission norm that resulted in a much better image than the existing ones. The usage of DBS allowed the EU to make a mandatory transmission norm: the MAC standard [8]. The scope of this Directive was limited in the sense that the standard was only mandatory if DBS were used for transmitting television signals. In case satellites that operated in the *fixed satellite services* (FSS) band were used for transmitting television signals the MAC standard did not have to be implemented. As a result, most satellite broadcasters started to operate from the typical telecommunications satellites (in the FSS band) and not from the DBS satellites. Hence, the success of this mandatory standard was very limited. Under the rules incorporated in the 1986 MAC Directive it was stipulated that five years after its implementation the effect of this Directive had to be evaluated. Because of the limited results of the first MAC Directive, the EU was reluctant to implement another standard that was not supported by the market and therefore tried to use pressure on the parties involved (manufacturers, broadcasters, and network operators) to sign a memorandum of understanding on how future television broadcasting was to be implemented. The memorandum was signed in the summer of 1992 under the condition that the EU would sufficiently subsidize the further technological development of HDTV. We now know that this did not happen.

The second MAC Directive of May 1992 [9] still chose HD-MAC and D2-MAC (*duo-binary multiplexed analog components*) wide-screen 16:9 aspect ratio but no longer made an analogous transmission norm mandatory. This second MAC Directive set a regulatory framework of standards on advanced television broadcasting services. In case of European satellite and cable transmission and digital HDTV that was not fully digital, television programs had to be based on the HD-MAC standard [10]. The D2-MAC standard [11] had to be used for not completely digital satellite and cable transmission with a wide-screen 16:9 aspect ratio format. Taking into consideration the developments in the United States, it opened the possibility of a digital transmission standard under the condition that such a standard would be an agreed standard by ETSI. The consequence of this approach was that the EU chose a European DTV system. This can also be concluded from the Council Decision of July 22, 1993, on an action plan for the introduction of advanced television services in Europe [12]. This action plan aims at promoting the wide-screen 16:9 format (625 or 1,250 lines), irrespective of the European television standard used and irrespective of the broadcasting mode (terrestrial, satellite, or cable).[2]

In the first half of 1993, it became clear to a number of governments, satellite operators, network operators, broadcasters, and manufacturers that a successful follow-up to the Eureka95 project within the context of the EU research and developments programs would be very unlikely. One has to compliment these parties on their initiative taken on May 29, 1993, to start the "European Launching Group of Digital Video Broadcasting," now known as DVB. The memorandum of understanding between the DVB parties was signed in September 1993. The European Commission was actively participating and wanted to cover the

---

2. Also the International Telecommunications Union (ITU) had already committed itself to the aspect ratio of 16:9 in ITU-R recommendation 709 defines *picture characteristics* including the wide-screen 16:9 aspect ratio.

developments by taking regulatory steps that would be acceptable to all parties, instead of making rules that were not acceptable to important parties in the market, as happened with the MAC Directives. The number of parties involved in DVB and the results of all earlier agreements within the Eureka95 project made rapid progress possible. By mid 1995 most (transmission) standards were agreed upon within the realm of DVB and forwarded to ETSI to make them official European standards. These results gave the EU the "courage" to propose a new ruling to cover the DVB developments. In this respect, it is interesting to cite some of the considerations in the preamble of the Television Standards Directive [13].[3]

... Whereas, for the purposes of this Directive, a wide-screen television service must meet the minimum requirements that it uses a transmission system delivering sufficient information to allow a dedicated receiver to display a full frame picture with full vertical resolution and whereas, for the same purposes, a television service using letterbox transmission in 4:3 frame which does not meet the above mentioned minimum criterion should not be considered as a wide-screen television service;

Whereas television services are currently delivered to the home by terrestrial systems, by satellite systems and by cable systems and it is essential that advanced wide-screen services should be made available to the largest possible number of viewers;

Whereas cable TV networks and their technical capabilities as defined by the Member States are an important feature in the television infrastructure of many Member States and will be crucial to the future of advanced television services;

Whereas master antenna systems as defined by Member States are not affected by this Directive;

Whereas it is essential to establish common standards for the digital transmission of television signals whether by cable or by satellite or by terrestrial means as an enabling element for effective free-market competition and this is best achieved by mandating a

---

3. In the literature, this Directive is sometimes referred to as the third MAC Directive, which is only relevant in case analog transmission is still being used. For digital transmission, there is, of course, no reference to the MAC transmission norm anymore.

recognized European standardization body taking account, as appropriate, of the outcome of the consensus processes under way among market parties;

Whereas such standards should be drawn up in good time, before the introduction to the market of services linked to digital television; ...

Important in the Television Standards Directive is the repeal of the second MAC Directive 92/38. With the completion of the most relevant agreed on (pre)standards within DVB and the EU regulatory coverage of these developments through the Television Standards Directive it can be concluded that real full-fledged digital HDTV was no longer the goal. Instead of a purely technologically driven development, the EU now aimed at a more market-driven and realistic approach to the implementation of a normal digital wide-screen television.

6.3.3   The current regulatory EU wide-screen TV-package

DTV services are considered to form an integral part of converging industries and sectors and, as such, are part of general (and sometimes specialized) laws and regulations. This section discusses the specific European regulatory framework on DVB.

6.3.3.1   EU Directive on television standards

DVB has proven to be able to provide the digital technology for television viewers. This was also recognized by the governments. Through the cooperation of national governments and the European Commission, the Television Standards Directive was accomplished. On July 25, 1995, the European Union's Council of Ministers, unanimously approved the Directive and established the Directive on October 24 of that same year. This Directive stated, among others things, that the second MAC Directive (92/38/EEC) was to be withdrawn nine months later. However, according to the Television Standards Directive, the D2-MAC standard, or any other system that is fully compatible with PAL or SECAM, still has to be used for the analog transmission of wide-screen (16:9) television programs. In case of the analog transmission of HDTV programs, the HD-MAC standard has to be used.

With regard to DTV broadcasting, the specifications developed in the DVB project, and made into standards by ETSI, were made obligatory. This namely concerns the norms for generating program signals and the adaptation of the transmission media of satellite, cable, and terrestrial networks. This generates an important political embodiment for digital wide-screen television.

The European Commission also intends to structure the market for DTV services, such as pay-TV, by using this Directive. For example, for the encryption of television programs the use of the *common scrambling algorithm* is mandatory. There are also stipulations that aim to realize a level playing field on the grounds of the Community competition rules. In fact, the opposing of dominant positions on the market is clearly stated.

Furthermore, digital (wide-screen) TV-sets must be provided with the *common interface*, which is not mandatory for the set-top box. Also, the Directive regulates the fair, reasonable, and non-discriminatory access to networks for DTV services. This restricts the positions for, among others, BskyB and Canal Plus. Furthermore, CA systems that are exploited on the market must dispose of the necessary technical possibilities for an inexpensive conveyance of control to the cable head-ends. Herewith, the CATV operators must be able to have complete control on a local or regional level over the services that use such systems for CA.

In addition, licenses concerning DVB specifications must be issued on grounds of non-discriminatory bases, so that no threshold arises for new parties coming onto the market. Subsequently, the member states must provide for arbitration procedures to settle unsolved disputed honorably, timely, and transparently.

The Directive had to be implemented by the Member States no later than August 24, 1996. However, not all member states adjusted their laws to this Directive in time. In Spain, this resulted in a dispute between *Canal Plus and Société Europénne de Controle d'Accès* (Canal+/SECA) and the Spanish State. On June 7, 1997, Canal+/SECA filed a complaint with the European Commission that Spain was breaching Community law by prohibiting the distribution of digital set-top boxes used by Canal Plus. In reaction, on June 27, 1997 the European Commission opened infringement proceedings (IP 97/564) against Spain for (1) violating the treaty with regard to the free movement principles; (2) failing to notify the concerning law, which prohibits the distribution of Canal Plus' set-top boxes, to the Commission; and(3) infringing the Television Standards Directive.

The European Commission sent a reasoned opinion (IP 97/680) to Spain on July 23, 1997, requesting the removal of provisions that violate the treaty from contested Spanish law on television set-top-boxes. Consequently, on September 12, 1997, Spain submitted a modified law. As a result, the European Commission decided to end the proceedings on October 8, 1997.

6.3.3.2 (draft) EU Directive on the Legal Protection against Piracy

The key issue in providing information services via CA systems is to ensure that only authorized users (i.e., users with a valid contract) can get access to a particular programming package. Encryption is often used as a means to technically protect these services from unauthorized access (piracy). In some Member States the legal protection against piracy is insufficient or even absent. This could lead to a widespread use of illicit devices (i.e., equipment or software designed or adapted to give access to a protected service in an intelligible form without the service provider's authorization). DVB recognized this problem from its beginning by producing recommendations [14] in October 1995 for the necessary flanking of the pan-European policy to discourage piracy. The European Parliament shared DVB's concerns and added a recital to the Television Standards Directive in order to establish an effective Community legal framework on antipiracy. This recital was adopted by the Council when establishing the Directive.

In March 1996 the European Commission published the Green paper "Legal Protection of Encrypted Services in the Internal Market" [15]. This Green paper discusses the regulatory measures that are required to protect services against piracy. Its preceding wide-ranging consultation confirmed the need for a Community legal instrument ensuring the legal protection of all those services whose remuneration relies on CA. As a result, the European Parliament and the Council proposed a Directive on the "Legal Protection of Services Based on, or Consisting of, Conditional Access" [16]. The current (May 1998) draft Directive provides for an equivalent level of protection between Member States relating to commercial activities that concern illicit devices. However, this draft Directive's implementation may not result in obstacles in the internal market concerning the free movement of services and goods.

The protected services concern radio, television, and information society services (e.g., video-on-demand, games, and interactive teleshopping) and the provision of CA to these services as a service in its own right. As such, this draft Directive prohibits and sanctions the manufacture, import, distribution, sale, rental, possession, installation, maintenance, or replacement for commercial purposes of illicit devices. Moreover, the use of commercial communications to promote illicit devices is prohibited and sanctioned. The draft Directive explicitly does not cover the private possession of illicit devices, intellectual property rights, the protection of minors, and/or national policies on the protection of public order or national security. The latter implicates that lawful interception as part of a national policy on cryptography is not considered piracy.

According to the DVB members, this proposal does not go far enough. DVB would like *"personal use and possession"* to be covered by this Directive as well. Moreover, DVB would like tougher sanctions to be imposed. As can be concluded from the current proposal, the European Commission and the Member States, however, did not adopt this idea. The final conclusions will be drawn as soon as the European Parliament, together with the Council, has established the concerned Directive.

6.4 Summary and conclusions

Europe has come a long way in its efforts to establish advanced television. The first attempts, based on a technology-driven approach to establish a European HDTV standard, resulted in failure. The market-driven DVB project proved to be more successful by providing for normal digital wide-screen (16:9) television. The strength of this approach is that DVB's market-driven specifications were turned into official standards by ETSI and that, next, the most important standards' use was made mandatory through the Television Standards Directive. Hence, the EU aimed at a more realistic and powerful approach toward the establishment of (now completely digital) advanced television services in Europe.

As already stated by the European Commission in its Green paper on convergence, the application of digital technologies drives the convergence of traditionally separated industries. Furthermore, multimedia services will increasingly be delivered via different infrastructures. Hence, the traditional boundaries between telecommunications and broadcasting are fading. It is this historical and technological background and that of telecommunications and competition law, against which the policy and regulatory environment for DTV services is created. In particular, this concerns the above mentioned Television Standards Directive, as well the (draft) Directive on the Legal Protection against Piracy.

By also providing for interactive DTV services, DVB in fact, entered the telecommunications (policy and regulatory) domain. However, as of now (1998) it remains to be seen whether the policy and regulatory developments in the area of telecommunications will, in time, also fully incorporate the future developments of DTV services in Europe. In this respect, broadcasting could even be considered as a form of telecommunications. It also remains to be seen whether these developments will lead to a completely liberalized television services environment. It is already possible to discern that PST will remain an important cornerstone of European public (cultural) policy. Separate public funding structures will either remain in place or will be recreated to enable public policy influence in this area. It is expected that the Green paper on convergence will play an important role in the European Union policy and regulatory environment to be developed for shaping the ICT revolution and, hence, the information society.

References

[1]   European Commission, *Green paper on the convergence of the telecommunications, media and information technology sectors, and the implications for regulation; towards an information society approach*, COM(97)623, 3 December 1997,p. 2.

[2]   European Commission, *Green paper on the development of the common market for telecommunications services and equipment*, Brussels COM(87) 290 final, 30 June 1987.

[3]   Council Resolution 93/C231/1 of July 22, 1993, on the review of the situation in the telecommunications sector and the need for further development in that market, 6 August 1993.

[4]   European Commission, *Green paper on a common approach to mobile and personal communications in the European Union; towards the personal communications environment*, COM(94) 145 final, 27 April 1994.

[5] European Commission, *Green paper on the liberalization of telecommunications infrastructure and cable TV networks—Part I: Principle and Timetable,* COM (94) 440, 25 October 1994.

[6] European Commission, *Green paper on the liberalization of telecommunications infrastructure and cable TV networks—Part II: A common approach to the provision of infrastructure in the European Union,* COM (95)682, 25 January 1995.

[7] van Eupen, Th. A. G., "Historisch overzicht," in *Handboek HDTV,* Kluwer Deventer, 1993, p. 10.

[8] Council Directive 86/529/EEC of 3 November 1986 on the establishment of common technical specifications on the MAC/packet standards family for direct satellite broadcasting, OJ No. L 311, 6 November 1986, pp. 28–29.

[9] Council Directive 92/38/EEC of 11 May 1992 on the adoption of standards for satellite broadcasting of television signals, OJ No. L 137, 20 May 1992, p. 17.

[10] ETSI standard reference: ETS 300 352.

[11] ETSI standard reference: ETS 300 250.

[12] Council Decision 93/424/EEC of 22 July 1993 on an action plan for the introduction of advanced television services in Europe, OJ No. L 196, 5 August 1993, p. 48.

[13] Directive 95/47/EC of the European Parliament and of the Council of 24 October 1995 on the use of standards for the transmission of television signals, OJ No. L 137, 23 November 1995, p. 17.

[14] DVB, *Recommendations on Antipiracy Legislation for Digital Video Broadcasting,* A006 rev 1, October 1995.

[15] European Commission, *Green paper on the Legal Protection of Encrypted Services in the Internal Market,* Brussels, COM (96) 76 final, 6 March 1996.

[16] Proposal for a European Parliament and Council Directive on the Legal Protection of Services Based on, or Consisting of, Conditional Access, 23 February 1998.

Note: All legal and relevant policy documents of the EU can be found at: http://www.ispo.cec.be

Analytical model

7.1 Introduction

As described in Chapter 3, technological developments in information and communications technologies and the individualization in modern societies are incentives to develop interactive multimedia services. DTV systems are ideal for providing these types of services. As such, they embody an important part of the convergence process.

Technological developments in the field of DTV have implications for society. Government policy makers must assess these developments and influence them if necessary. The same is true for decision makers in industry and consumer organizations, as they also need to develop strategies based on their various interests.

This chapter discusses several aspects that play a role in the provision of interactive DTV services via CA systems and presents a conceptual model to illustrate the relationship between the technological developments and their consequences. Moreover, this model visualizes per consequence what needs to be considered in order to responsibly incorporate the concerned technologies into society.

7.2 Technological development aspects

This section discusses several aspects of the technological developments in the field of interactive DTV services, which are provided via CA systems [1].

7.2.1 Availability

In the information society there are services with a great social and economic interest. Examples are telephony, telex, and telegraph services and national television broadcasting. The availability of these services must be guaranteed. It may be that in the (near) future certain other services, which are provided via CA systems, become of great social and economic interest. For example, elections or referenda can be facilitated by CA systems. A user can insert a smart card with a pin-code in a set-top box's smart card reader for identification, after which authorization can be given to cast a vote.

7.2.2 Multiformity

In a situation in which only one or a few information service providers are active on the market, there is a considerable dependence on the supply of the service provider(s). It may occur that only a limited number and/or a one-sided type of service is available. A great variety of information, from various sources and from different perspectives, is of great interest for society. In other words, the multiformity of information is an important social requirement.

7.2.3 Affordability

The price consumers have to pay for their services may be too high for certain groups in society. This may lead to a situation in which society is divided into *haves* and *have nots*. A scenario in which some people are financially unable to access services that are considered necessary for social functioning (e.g., telephony and national television broadcasting) must be prevented. The affordability of these services must be guaranteed.

In 1997 a big debate took place in the Netherlands on whether it should be possible to exclusively broadcast soccer matches (which are very popular) via a pay-TV channel. These matches had always been broadcast via free-TV channels, including national television. However, a new pay-TV consortium called *Sport7* managed to produce the highest bid on the broadcasting rights for the most important (inter)national soccer matches. People would now be forced to pay extra for a set-top box and a subscription fee. Even the prime minister himself was involved in the discussions. In the end, the discussions ended when it turned out that the consortium was unable to finance the broadcasting rights to the matches. As a result, an established pay-TV broadcaster bought the rights to the national competition's live broadcasts. The national competition's summaries and the international matches are still broadcast via free-TV channels.

7.2.4 Market structure

The characteristic feature of open CA systems is that other providers' set-top boxes can be used. This is not possible with closed systems by which means market positions can be protected and even (regional) monopolies can be created. This is currently the case with most cable, satellite, and terrestrial systems. At this moment, large broadcasters (e.g., BskyB and Canal Plus) are using proprietary systems, which more or less have become de facto standards.

Broadcasters and CA system manufacturers are mutually dependent. It is undesirable for broadcasters to require manufacturers to construct television sets or set-top boxes that exclude other systems (i.e., the possibility to access services from other service providers). This would result in a closed system.

If there were to be a completely open standard without any control on the issue of licenses and if the royalties would be acceptable, then everybody would be able to implement a CA system via this standard. This would mean that every licensee could obtain the algorithm, the required cryptographic keys, and the means to construct various system security and control applications. This requires the system's information to remain strictly secret. In case this information becomes public, everybody, including pirates and hackers, could control the CA system and obtain free access to the services. The systems would be open but not serviceable.

Alternatively, an organization that manages the entire CA system could be created. This idea would probably lead to resistance from the broadcasters, as they would be dependent on another organization for their income. Also, this organization, based on its consolidated technical knowledge, could decide what equipment would be tested and used and from which supplier the equipment would be bought. In case this organization happened to be the only supplier, a monopoly could arise. Moreover, other conflicts regarding new features and the replacement of compromised systems could arise.

The best scenario that can be achieved seems to stem from the use of an open set-top box. The broadcasters control their proprietary CA Management System (CAMS). This requires all set-top boxes to be based on the same specifications. The verification of the user's identity and authorization takes place via a separate proprietary module (e.g., a smart card). Hence, the open set-top box has to be equipped with a common crypto system (also called *common scrambling algorithm* by DVB) and a common interface between the proprietary module (i.e., the CAMS) and the common crypto system. The use of this interface implies a cost increase. The set-top box specifications have to be administered by an organization that is specially assigned to this task. This allows free competition between set-top box manufacturers on the condition that it must be commercially viable for both the manufacturers and the broadcasters.

7.2.5 One-stop shop

In the current situation, broadcasters administer their proprietary set-top box population and take care of the CA system's management (i.e., billing and subscriber authorization). The consumer has to turn to the broadcaster for his or her subscription. Next, the consumer obtains a set-top box, by which means the broadcaster's encrypted programming package can be received and decrypted.

The advantage of the open set-top box is that the broadcaster can still use its proprietary CAMS. However, the user now has to sign up with different broadcasters in order to access the various services via his or her open set-top box. Moreover, consumers might need different modules from the various broadcasters. This could block the entry of new broadcasters into the market and, hence, hinder the establishment of a competitive market.

From the consumer's perspective it would be useful to create one counter through which all services from the various broadcasters could be provided. The counter functions as a one-stop shop. This has three important advantages. First, the consumer does not have to sign up with every single service provider. Next, the consumer does not need different modules to access the various services. Finally, competition takes place on the quality of service, rather than on the access to networks.

7.2.6 Privacy

With the provisioning of information services, CAMSs can be used to register personal data. The service providers need this data for their billing process. Moreover, personal data is often used internally to statistically analyze the market. The personal data obtained via interactive services are rather sensitive, since users' names, addresses, and residences are linked to their individual consumptive behavior. This kind of information can be used to construct a consumer profile. The fear that personal data is passed on (i.e., sold) to third parties (e.g., direct marketing organizations) is justified. For example, there are *sucker lists* that indicate which individuals have decided to buy a certain product immediately after it has been shown in a commercial. These lists are valuable information for direct marketing organizations.

7.2.7 Cost allocation

Investments are required to adjust a network to facilitate interactive television services, which are provided via CA systems. The location as well as the level of costs depend on the choice for a specific design. A design can introduce additional or less costs for actors in other layers within the layer model. Ideally, the costs are divided over the different layers in such a way that the service provision is cost-efficient. In the end, this leads to lower costs for the consumer, which in turn allows the market to grow.

As stated earlier in Section 7.2.4, standardization in the field of CA may result in an open set-top box. This implies that, in contrast with a closed system, actors other than the CA service provider obtain options from which they can choose. Hence, it is possible that certain costs are shifted to other layers. For example, the programming package's encryption is currently processed by the information service providers (i.e., pay-TV broadcasters) themselves. It could very well be that in the future network service providers will provide the encryption for all (new) information service providers. This means that new information service providers do not have to invest in such a crypto system. Another example concerns the multiplexing process. In case multiplexing is applied centrally by the network service provider, this is more cost-efficient than if this were to be done by each information service provider independently.

These examples illustrate that costs can and probably will shift to different layers. If this results in a concentration of costs within one single layer (i.e., one group of actors), a threshold will arise to invest in these new technologies. Hence, innovation is delayed or does not even take place at all.

7.2.8 Lawful interception

The function of CA systems is to ensure that only authorized users (i.e., users with a valid contract) can watch a particular programming package. Technically, this means that a television program is broadcast in encrypted form and can only be decrypted by means of a set-top box. The set-top box incorporates the necessary hardware, software and interfaces to select, receive, and decrypt the programs.

It is very important that the crypto system be unable to be compromised so that services can be accessed for free. This requirement is even more important if the crypto system is standardized, because if this system is compromised, free access can be obtained on a larger scale than with proprietary systems. In the latter case, every system has to be compromised separately.

The need for information to be encrypted, however, conflicts with the national security and law enforcement agencies. These agencies have the need to lawfully intercept telecommunications traffic, for example, in order to anticipate terrorist attacks or criminal activities. If this traffic is encrypted by means of the CA system's crypto system, the intercepted data is unintelligible. Hence, the concerned agency is not able to anticipate the communicated information. From this perspective, the crypto system may not be too complex.

7.2.9 Intellectual property rights

Television programs are subject to intellectual property rights. Broadcasters have to pay the concerned information producer for the (exclusive) right to broadcast a particular program. Sometimes very high prices have to be paid, especially in the case of single events (e.g., boxing matches). In the case of free-TV, the broadcaster finances its activities by, among other things, including commercials in its programming package. A pay-TV broadcaster often pays a higher price for broadcasting rights, because the program has not yet been shown to a large audience—except in the case of movies, which are shown in the cinema prior to television. Moreover, most pay-TV broadcasters do not include commercials in their programming package. Hence, the price of pay-TV programs is relatively high. This and the fact that pay-TV broadcasters often show the latest movies before they become available on a free-TV channel, sometimes lead to creative solutions in trying to watch these programs for free.

In this context, piracy refers to the unauthorized access to services that are provided via CA systems. It also applies to the situation in which legal hardware and/or software is used. The actors who try to get access without authorization are called pirates. In general, CA systems are not compromised through breaking the crypto system. Unauthorized access is often obtained through the features, such as access to services via a free subscription, extension of the program's first free three minutes, and last, but not least, the interruption of messages to the set-top box. It is self-evident that piracy undermines investments (i.e., innovation) in this field.

7.3 Conceptual model

An open market structure for interactive television services, which are provided via CA systems, can optimally be achieved by the application of an open set-top box. This set-top box has to incorporate a common scrambling algorithm and a common interface between this common scrambling algorithm and the proprietary module (i.e., CAMS).

By means of a conceptual model (see Figure 7.1) the relationships between the technical developments (indicated by circles) and the consequences (represented by square boxes) of these developments are described. Moreover, this model identifies per consequence what aspects need to be taken into account in order to responsibly incorporate these technologies into society. The aspects, in turn, are positioned below the square boxes.

For society, it is important that:

▶ Services with a great social and economic interest are available;

▶ The information's multiformity is assured in the case of a limited and one-sided supply of services;

▶ The affordability of basic services is guaranteed if high costs lead to haves and have nots;

▶ An efficient and effective cost allocation in the economic value-added chain is achieved in case costs are shifting to different layers in the economic value-added chain;

▶ An open market structure is established despite the cost increase due to the application of a common interface;

▶ A one-stop shop is created in the situations where many CAMSs (i.e., modules) are used;

▶ Privacy is protected in case personal data is registered;

▶ If strong crypto systems are used, lawful interception is assured in order to protect national security and public order. On the other hand, the crypto system must be strong in order to protect intellectual property rights against piracy. Lawful interception implicates that privacy protection is limited;

▶ Intellectual property rights are (legally) protected, in case a CA system is subject to piracy.

7.4 Summary and conclusions

Ideally, all factors are considered in such a way that interactive DTV services, which are provided via CA systems, are responsibly embedded in

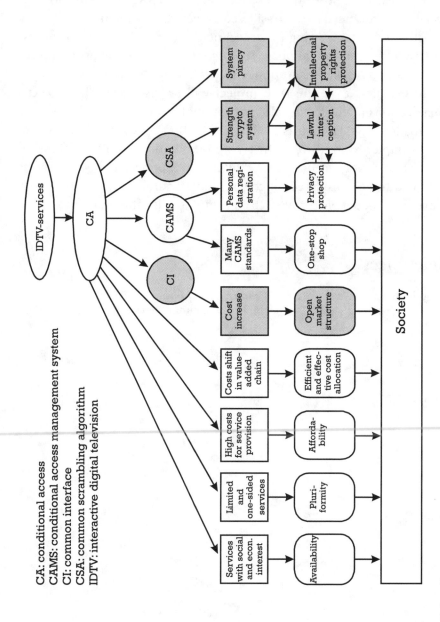

CA: conditional access
CAMS: conditional access management system
CI: common interface
CSA: common scrambling algorithm
IDTV: interactive digital television

Figure 7.1 Conceptual model.

society. In the first instance, the market parties should address these aspects. Typically, market parties focus on the technological and/or eco- nomical aspects first. However, the aspects with a social and/or institu- tional character need to be considered as well. Governments should play an active role by assessing these technological developments. If neces- sary, governments can influence them, so that not only techno-economic innovation, but socio-institutional innovation is achieved as well.

In Chapters 8-12 the DVB project will be discussed. The project's organization and the technical specifications for several (interactive) digi- tal broadcasting systems, as well as the DVB CA system, are described. Next, Chapter 13 concerning the DVB project's analysis will explain what aspects have (not) been addressed by the DVB project and what aspects still need to be considered by governments and/or market parties. The conceptual model will prove to be a useful tool for this analysis.

References

[1]   de Bruin, R., *Technologie Beleidsonderzoek naar Interactieve Digitale Video-diensten met Conditional Access*, Technische Universiteit Eindhoven, October, 1995.

CHAPTER

8

Contents

European digital video broadcasting project

8.1   Introduction

As stated in Chapter 1, the government-driven approach to establish a European standard for an analog satellite HDTV broadcasting system appeared to be a failure. The HDMAC Directive, which was meant to set a standard, was abandoned as a policy line. Meanwhile, in the United States, the FCC charged the GA with developing a standard for a digital terrestrial HDTV broadcasting system. In reaction, market parties in Europe initiated the market-driven DVB project for the development of digital wide-screen (16:9) television.

This chapter discusses the DVB project's developments, first providing the background and subsequently explaining the project's structure and related research projects. Next, the standardization process, including the European Commission's and the national governments' roles, is discussed. This is

125

followed by a description of the various cooperating standardization bodies and groups. Finally, the ambitious planning and the results are described.

8.2   Background

The European DVB project's technical basis was formed in late 1990. In an experimental European project called SPECTRE, it was proved that it is possible to effectively reduce the transmission capacity that is required for DTV. Until that point in time, it was not clear whether it was possible to practically implement digital coding systems. The concerned compression system was also known as the motion compensated hybrid discrete cosine transform coding system.

In reaction to the developments in the United States, the Scandinavian *HD-DIVINE* project on the development of an HDTV standard for digital terrestrial broadcasting started in 1991. Moreover, Swedish television launched the idea of a pan-European platform for European broadcasters, with the objective of developing digital terrestrial broadcasting. Meanwhile, in Germany conversations took place concerning a feasibility study on current television technologies and the alternatives for the development of television in Europe. In late 1991, the German government recognized the strategic importance of DTV in Europe and the need for a common approach and invited broadcasters, telecommunication organizations, manufacturers, and regulatory authorities in the field of radio communications to an initial meeting that led to the formation of the ELG in the spring of 1992. The ELG expanded and on September 10, 1993, a memorandum of understanding was signed by 84 European broadcasters, telecommunication organizations, manufacturers, and regulatory authorities. Together, they now formed DVB. In the spring of 1998, the DVB project counted more than 200 members from 30 countries worldwide. The memorandum of understanding contained the rules with which members had to comply, based on their common interest and mutual respect. The three main objectives were the following:

- The promotion of and contribution to the definition of technical standards on DTV and their widespread use and application;

▶ The facilitation of the introduction of new services that use these standards, including research on related fields such as frequency planning and CA;

▶ The facilitation of the closest possible coordination between precompetitive research and development and standardization.

Research of the working group on DTV in the field of terrestrial DTV introduced several new important concepts, among which were proposals to simultaneously serve different consumer markets. Moreover, it became clear that the MAC systems for HDTV satellite broadcasting were to be overtaken by digital systems. At that moment DVB provided the ideal platform for a common approach by all parties concerned in order to develop a completely DTV system.

At the same time the European market demanded more channels, rather than a system with a better performance such as HDTV. The application of compression techniques on digital signals allows a dramatic bandwidth reduction, so that more channels can be created within the same available bandwidth. An HDTV signal, however, requires more bandwidth than a normal TV signal. This also applies to the digital domain. Furthermore, digital transmission allows the application of forward error correction, which results in a better display quality. Hence, DVB aimed at normal digital wide-screen (16:9) television, rather than digital HDTV.

It was recognized soon that the development of digital cable and satellite television systems had to be started first, as they presented fewer technological and legislative problems than terrestrial systems. The market also demanded that these systems become a priority.

8.3   Project structure

The memorandum of understanding signatories are members of the General Assembly, which meets on an annual basis. This is the highest decision-making forum within the DVB project. The General Assembly has elected a Steering Board, which is small enough to efficiently make decisions and large enough to represent all DVB members' different interests. The Steering Board, which is DVB's executive committee, is supported by approximately 20 subgroups. It was agreed that the steering board should represent four different interest groups: broadcasters, manufacturers, telecommunication organizations, and regulatory bodies. The Steering Board's seats were assigned as follows:

▶ Twelve broadcasters;

▶ Eight manufacturers;

▶ Eight telecommunications and satellite organizations;

▶ Six regulatory bodies;

▶ Representatives from ETSI, the Comité Européen de Normalisation ELECtrotechnique (CENELEC), and the European Commission were admitted to the Steering Board as observers.

Meanwhile, there are four modules that report to the Steering Board. These are the technical module and the commercial modules for cable and satellite, terrestrial, and interactive services. Below these modules a large number of subgroups are occupied with detailed (technical) designs. The actual work on the technical designs is carried out by ad-hoc working groups and a special reporting group that, in turn, report to the concerned modules. The ad-hoc group members are specialists from organizations that are involved in the DVB project. These specialists are occupied in the following fields:

▶ CA;

▶ Regulatory aspects;

▶ Budgeting;

▶ Procedure rules;

▶ Promotion and communication;

▶ Intellectual property rights.

Figure 8.1 presents the DVB organizational structure [1].

The ad-hoc group on CA has worked on the specification of one CA system in which the following aspects, among others, play an important role:

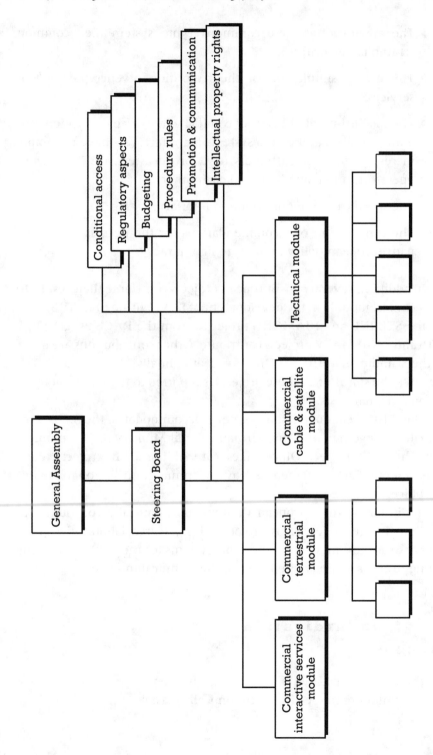

Figure 8.1 DVB organizational structure.

▶ The application of one common crypto system (i.e., common scrambling algorithm);

▶ The use of simulcrypt for the realization of encrypted video-services;

▶ The establishment of a code of conduct with which encryption program suppliers and crypto systems manufacturers have to comply in respect to their customers (e.g., broadcasters, CATV-network operators, other suppliers, and each other);

▶ The specification of one common interface;

▶ The formulation of recommendations for the necessary flanking pan-European policy in order to fight piracy.

In addition, several pan-European projects contribute their results to the technical module in the development of DTV. Within SPECTRE these are the STERNE and DIAMOND projects. From the RACE program the dTTb and DIGISMATV projects contribute. Other contributions are made by the Scandinavian HD-DIVINE project and, finally, the German HDTV projects. The status of all these projects is reported to the technical module on a regular basis.

The EBU headquarters in Geneva accommodates the DVB Project Office. Together with the German Federal Ministry of Communication, the DVB Project Office takes care of the DVB project's daily management. Each DVB member pays an annual contribution to finance this office.

Additionally, the European Commission financially contributes to the Euro Image Project. This project helps research laboratories cover their costs for testing satellite and cable systems for the DVB project. The testing results are directly reported to the technical module.

8.4 Standardization process

The three commercial modules formulate user requirements, which have to be met in order to specify an economically feasible system. Next, these requirements have to be translated into technical specifications by the technical module. The procedure requires that the concerned commercial module has to approve the technical specification first, before its final approval by the Steering Board. After the Steering Board's approval, these technical specifications are submitted to the relevant standards body (i.e., ETSI or CENELEC). After adopting the technical specifications, these bodies turn them into official standards.

Next, the Directive on television Standards [2] obligates the use of several standards on DTV and CA from officially recognized standardization bodies. This forms a very important incentive for parties to participate in the DVB project, as these parties' interest is to influence the development of (obligatory) standards. Another important incentive is the establishment of a European and preferably a world standard, because the national European markets are often too small to achieve economies of scale.

8.5 Related standardization bodies and groups

Beside the DVB project's members, various standardization bodies and other groups are involved in the DVB project. This section describes the role of these organizations and groups.

8.5.1 International Telecommunications Union (ITU)

The ITU is the most important standardization organization at a global level in the field of telecommunications. The ITU can be divided in three sectors: the radio communication sector (ITU-R), the telecommunications standardization sector (ITU-T), and the development sector (ITU-D). Within the ITU-R, two working parties are active: Working Party 10-11/S in the field of satellite systems and Working Party 11/3, which is occupied with digital terrestrial systems. The latter has appointed a special reporter to study the future developments of a common digital terrestrial system, or common parts thereof. The ITU-T has appointed a special reporter as well.

8.5.2 ISO/IEC

The *International Standardisation Organisation* (ISO) and the *International Electrotechnical Commission* (IEC) are working at a global level on the standardization of consumer and industrial equipment. The ISO is a general standardization organization, while the IEC focuses on the standardization of electronic equipment. Due to the overlap between both organizations in the field of information technology, the *Joint Technical Committee 1* (JTC1) was established. JTC1 is charged with standardizing information technology-related equipment.

A JTC1 sub-group called MPEG has developed a standard for baseband video compression and a multiplex system for video with VHS quality (MPEG-1) and audio with CD quality. Next, the high-quality MPEG-2 standard was developed. Due to its flexible and compression character, this standard plays an important role in DTV broadcasting.

8.5.3 Comité Européen de Normalisation Electrotechnique (CENELEC)

The CENELEC is working at a European level on the standardization of consumer and industrial equipment. This organization incorporates technical committees in the field of television, radio receivers, CA, and cable distribution systems. In the past, the CENELEC has standardized a CA system for MAC/packet services. Hence, it is a suitable organization to contribute to the standardization of DTV.

8.5.4 European Broadcasting Union (EBU)

The EBU is an organization of European public broadcasters. Non-European broadcasters can join the EBU as *associate members*. The EBU contains about 50 members and more than 60 associate members from all over the world. The EBU establishes and publishes recommendations and standards, which are often considered by the ITU and/or the IEC to be turned into world standards. With respect to the DVB project, the EBU substantially contributes to drafting system requirements, system evaluations, and frequency planning.

8.5.5 EBU/ETSI JTC and EBU/CENELEC/ETSI JTC

ETSI (European Telecommunications Standards Institute) was established by the European Commission to develop standards that can be implemented in the member states by means of regulations (i.e., Directives). Parties from all sectors, including regulatory bodies, manufacturers, and broadcasters, are allowed to participate in ETSI. The EBU/ETSI JTC (Japan Telecom) is responsible for standards on broadcast signals and point-to-point transmission of broadcast signals (i.e., transmission-related standards). This committee reports to both the ETSI technical assembly and the EBU technical committee. The formal approval for each standard on DTV is processed via the EBU/ETSI JTC according to the ETSI procedures. The EBU/CENELEC/ETSI JTC is responsible for the guidelines and standards concerning source coding and multiplexing, CA, and interactive services.

8.5.6 DAVIC

January 1994 marked the establishment of the Digital Audio-Video Council (DAVIC), which has more than 100 members worldwide, among which is the European Commission's directorate general XIII B. DAVIC promotes a common vision on a digital audio-visual world, in which producers of digital audio-visual programs can reach as wide an audience as possible, users have equal access to services, network service providers can effectively carry out the transport, and manufacturers can supply hardware and software for a free production of information, the supply of information streams, and the use of the information itself.

Additionally, DAVIC tries to stimulate the introduction of interactive services by providing the appropriate international specifications for open interfaces and protocols. Moreover, DAVIC submits these specifications to the relevant international standardization organizations and cooperates with these organizations in the standardization process. If the required specification does not exist, DAVIC contributes to its development. In principle, the members' use of these specifications is voluntary.

8.6 Results DVB project

To prevent the DVB project's takeover by new technological developments, it was necessary to plan tightly. Figure 8.2 illustrates the strict program that comprised the DVB project's beginning phase.

The DVB project aims to develop specifications on DTV. The most important DVB specifications are listed in Table 8.1, which also includes their application, their official ETSI standard definition, the date on which they became official standards, and their latest version (status as of December 22, 1997).

Beside the specifications listed above, DVB has produced several guidelines, which cover, among other things, the use of MPEG-2 audio- and video-source coding and multiplexing. Currently, the specifications of interfaces to *plesiochronous digital hierarchy* (PDH), *synchronous digital hierarchy* (SDH) and *asynchronous transfer mode* (ATM) networks are in the final stage of approval in ETSI. Moreover, a *multimedia home platform* (MHP) is being developed. The MHP forms the *application protocol interface*(API) to all different kinds of multimedia applications. Finally, DVB is working on the transmission of data in DVB bitstreams. This allows operators to, for example, download software over satellite, cable, or terrestrial links, to deliver Internet services over broadcast channels (using IP tunneling) or to provide interactive television. For this purpose the MPEG-2 DSM-CC (Digital Storage Media— Command and Control) standard is used to provide the data broadcasting system's core.

As described in Section 8.5, DAVIC is involved in the DVB project. DAVIC decided to adopt the DVB transmission specifications. The ITU has adopted the DVB specifications as well and formalized them in ITU Recommendations for DTV. It must be noted that no other standards on DTV were available at that time. Moreover, DVB and DAVIC have worked closely in developing specifications for interactive services. On March 18, 1998, ITU-T Study Group 9 approved a standard [3] on transmission systems for interactive cable television services. The standard includes three annexes addressing the various requirements of European (i.e., DVB), North American, and Japanese sectors.

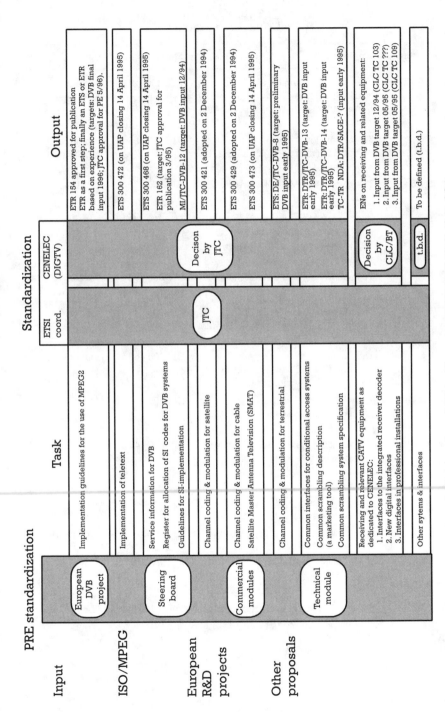

Figure 8.2 Planning DVB project in December 1994. (*Source:* EBU/ETSI JTC (94) 28.)

Table 8.1
Digital Television Standards

Specification	Application	Standard	Date
DVB-S	The satellite system for use in the 11/12-GHz band, suitable for transponders with various bandwidths and powers	ETS 300 421 EN 300 421	December 1994 August 1997
DVB-C	The system for CATV networks, compatible with DVB-S, for 8-MHz channels	ETS 300 429 pr EN 300 429	December 1994 to be published
DVB-CS	The system for SMATV for providing CAI	ETS 300 473 EN 300 473	May 1995 August 1997
DVB-T	The terrestrial television system, compatible with DVB-S and DVB-C, applicable to terrestrial 7-8 MHz channels	ETS 300 744 EN 300 744	February 1997 August 1997
DVB-MS	The MVDS for $f > 10$ GHz, compatible with DVB-S	ETS 300 748 EN 300 748	October 1996 August 1997
DVB-MC	The microwave multipoint distribution system for $f < 10$ GHz, compatible with DVB-C, for 8-MHz channels	ETS 300 749 EN 300 749	April 1997 August 1997
DVB-SI	The service information system for configuration and adjustment of the DVB set-top-box to DVB format bitstreams	ETS 300 468 prEN 300 468	January 1997 to be published
DVB-TXT	The teletext specification for the transport of standard teletext in DVB bitstreams	ETS 300 472 EN 300 472	May 1995 August 1997
DVB-SUB	The subtitling system for the display of graphical objects (e.g., subtitles and logos) on the television screen	ETS 300 743	September 1997
DVB-SIM	The technical specification of simulcrypt in DVB systems; part 1: head-end architecture and synchronization	TS101 197-1	June 1997
DVB-CI	The specification of the common interface for CA and other applications	EN 50221 (CENELEC)	February 1997
DVB-NIP	The specifications of network independent protocols for interactive services	ETS 300 802	November 1997
DVB-RCC	The specification of interaction channels through CATV networks	prETS 300 800	to be published
DVB-RCT	The specification of interaction channels through PSTN/ISDN	ETS 300 801	August 1997

8.7 Summary and conclusions

In contrast with earlier initiatives in Europe and the United States, the DVB project can be characterized as purely market-driven. Strict commercial requirements were established by market parties that participated in the DVB project. Working to tight time schedules and strict market requirements allows for the necessary economy of scale to be achieved. This ensures that, in the transformation of the industry to digital, broadcasters, manufacturers, and finally, the viewing public will benefit. The fact that ETSI turned these specifications into official standards proved to be a powerful second step. Next, the European Commission and the member states developed the Directive on Television Standards, which forms an important additional incentive for the establishment of international standards.

In addition to traditional broadcasting, DVB provides all kinds of additional services. This not only concerns broadcasting-related services such as, for example, system information and subtitling, but different types of interactive (multimedia) services as well. By providing for the latter, DVB, in fact, entered the telecommunications (policy and regulatory) domain. Hence, DVB has caused a paradigm shift by combining traditional broadcasting and telecommunication services.

The adoption of the DVB specifications on the various transmission systems by DAVIC and the ITU has considerably heightened the possibility that these DVB specifications will be recognized as world standards. Moreover, the annexing of the DVB specifications to the ITU standard on transmission systems for interactive cable television services can be considered a major milestone in the establishment of international standards for bidirectional digital cable systems. In view of these achievements, the DVB project can already be called very successful.

References

[1]   Smits, J., *DVB: fundament op weg naar Europese Elektronische Snelweg?*, in Kabeljaarboek, December, 1994.

[2]   Directive 95/47/EC of the European Parliament and of the Council of 24 October 1995 on the use of standards for the transmission of television signals, O.J. L281/51, 23 November, 1995.

[3]   ITU Recommendation J-112.

CHAPTER

9

Contents

Coding techniques and additional services

9.1   Introduction

The digitization of analog audio and video signals increases the signals' bandwidth. The application of source coding results in a decreased signal bandwidth, and several bit rates can be selected to support the required quality of service. Eventually, the digitally coded signals require less bandwidth than analog signals, with the same quality. The most common standards for the digital coding of audio and video signals are produced by the ISO/IEC JTC1 SC 29 MPEG. This group has specified and upgraded several standards for digital audio-visual coding. In the case of audio, the current standards are known as MPEG layer I to layer IV. For digital video coding, the current standards are referred to as MPEG-1 to MPEG-4.

This chapter discusses the DVB specifications for source coding and additional services. DVB has decided to make use of the

MPEG standards for source coding and the multiplexing of audio-visual signals and has thus produced guidelines (not specifications) for implementing MPEG-2 audio, video, and systems in satellite, cable, and terrestrial broadcast applications (ETR 154 [1]). These guidelines represent a minimum functionality that all *integrated receiver decoders* (IRDs) can either meet or exceed. The IRDs that meet these minimum functionalities are called baseline IRDs. All features other than those provided in the guidelines are left to the marketplace.

Additionally, DVB has produced specifications for service information, teletext systems, and subtitling systems. These specifications are standardized by ETSI. First, the basic elements of the systems mentioned above are discussed before explaining the specifications for several encoding and decoding processes at a functional level. Information concerning the performance of these systems can be obtained from the ISO/IEC and ETSI standards papers, as referred to in the concerning sections.

9.2   MPEG-2

The MPEG-2 standard consists of three parts. Two parts concern the audio and video (source) coding. The other part, referred to as MPEG-2 systems, provides a standard for the multiplexing of digital audio and video signals. Section 9.2.1 discusses the basic elements of digital coding.

9.2.1   Elements of digital coding

This section discusses how an analog signal is processed into a digital signal. To obtain a digital audio or video signal with less bandwidth than the original analog signal, the concerned analog signal is coded. Accordingly, Section 9.2.1.2 explains the coding techniques, which make use of the limitations of human perception.

9.2.1.1   Digitization of analog signals

The sounds of a television show (e.g., the actors' voices) are transformed into analog signals by means of a microphone. The same process applies to a camera, which transforms pictures into analog signals. These analog, audio, and video signals are typically continuous in the time domain.

An *analog-to-digital* (A/D) converter is used to obtain a digital representation of an analog signal. For the first step of this conversion process, a sample is taken from the analog signal at discrete points in time. According to the Nyquist theorem, the sample frequency (fs) has to be at least twice as high as the signal bandwidth (fs $\geq$ 2B [Hz]). Next, the signal amplitude of every sample is binary coded with k bits, which means $2k$ quantizing levels can be distinguished. Hence, a sequence of binary coded samples represents the analog signal.

Eventually, the digital signals are converted into analog signals again by a D/A-converter. After this conversion, the audio and video signals are transformed into sounds and pictures by loudspeaker(s) and the cathode-ray tube of the television set, respectively.

9.2.1.2 Digital coding and human perception

The objective of digital source coding is to achieve a better transmission quality and at the same time, as bandwidth is limited, to allow a reduction of bandwidth. The latter, which is also known as compression, can be achieved by reducing either redundant or irrelevant information. The redundancy of information is a measure of the signal predictability. The reduction of redundant information requires prior knowledge of the statistical characteristic of the signal. These redundant parts of information are then reduced by the encoder. At the decoder, the original signal is reconstructed by adding a substitute of the redundant information to the encoded signal. Additional information is inserted by the encoder to calculate the substitute. However, the predictability of audio signals, such as speech, is very low. Consequently, a very limited redundancy reduction can be achieved.

As a result of the limitations of human hearing with respect to amplitude, time, and frequency spectrum, part of the original information cannot be noticed and is therefore irrelevant. A psycho-acoustical model describes the hearing characteristics of the average human being. By means of this model, it is possible to mask irrelevant information and only encode relevant information. Hence, a smaller signal bandwidth is achieved. In contrast with redundancy reduction, the reduction of irrelevant information is irreversible. Besides, the irrelevant information cannot be picked up by the human ear, so there is no need to add a substitute of this information at the decoder.

Because human vision, just like human hearing, is limited in amplitude, time, and frequency, irrelevant information can be reduced to obtain a smaller signal bandwidth. Moreover, the predictability of video signals (e.g., in the case of static pictures) can be relatively high. This allows a significant redundancy reduction.

9.2.2 Audio coding

This section explains the relevant MPEG encoding and decoding standards and discusses the DVB guidelines for the use of these standards.

9.2.2.1 Audio encoding

The DVB guidelines for encoding audio signals are based on the ISO/IEC IS 1381-3 standard [2]. This standard defines the MPEG layer I and layer II coding. The latter provides a higher compression level with remaining audio quality but is more complex and costs more to implement. Audio signals coded with MPEG layer II approximates CD-quality and is frequently applied in other audio products worldwide. Both standards make use of the reduction of irrelevant information—i.e., sounds beyond the human hearing are not encoded. Figure 9.1 presents a functional description of the MPEG layer I audio encoding system.

The analog audio signal is divided into 32 subbands by a filter bank. The MPEG-2 standard describes the characteristics of this filter bank. Next, the signals within the subbands are digitized by a process of sampling and quantizing. The maximum amplitude of every 12 samples forms a *scale factor,* which is provided to the input of a psycho-acoustical model. By means of the scale factor, this model calculates the required (maximum) quantizing level for each of the 12 samples.

Additional input is provided to the psycho-acoustical model. Parallel to the division of the audio signal into 32 subbands, the Fourier transformation for successive parts of the audio signal is calculated. Each part consists of 512 samples. Subsequently, local maximums within the spectrum of the 512 samples are detected. The value of a local maximum is compared to samples near the local maximum's frequency. Through this process it is possible to see whether local maximums are part of the actual audio signal. If so, the psycho-acoustical model is adjusted to mask the local maximums as well. The shape of the "mask" together with the bit rate eventually define the optimal number of quantizing levels. This

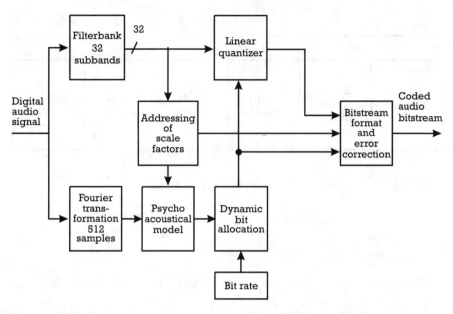

Figure 9.1 Functional description of the MPEG layer I audio encoding system (mono).

means that the number of bits per sample can vary, thereby allowing a dynamic reduction of the signal bandwidth.

Finally, the output of the quantizer, the scale factors, and the value of the number of bits per sample are processed into a bit stream format (see Figure 9.2). This format includes a header, which contains 12 synchronization bits and 20 bits for system information. Optionally, the value of the number of bits per sample and a part of the header can be protected by means of error correction data (16 bits). The bit stream format allows additional bits to be included. These bits can, for example, be used to upgrade the audio system with surround sound.

The MPEG layer II audio encoding system (see Figure 9.3) is compatible with the layer I system. The layer II system supports mono, stereo, multilingual sound, and surround sound. One of the main differences between the systems is that the layer II system calculates a scale factor for every 36 samples. When several spikes in the audio signal occur, one scale factor for 36 samples may not be sufficient. In this case, for every subband two or even three scale factors may be required. Depending on the character of the audio signal, the required number of scale factors can be

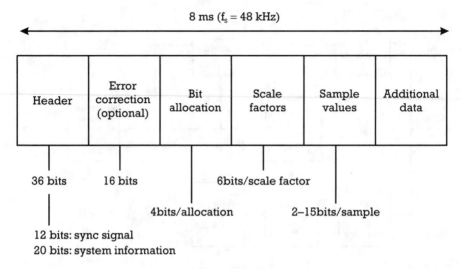

Figure 9.2   Bit stream format of the MPEG layer I system (mono).

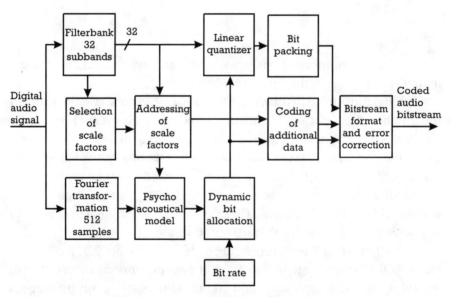

Figure 9.3   Functional description of the of MPEG layer II audio encoding system (mono).

selected. The value of the number of scale factors is encoded as additional information.

Another difference between the systems is that a reduced number of bits per sample (two or three bits instead of four bits) is used for subbands located at higher frequencies. In general, the signal energy is considerably lower at higher frequencies. Hence, less quantizing levels (and thus less bits) are required to represent that part of the signal. As a result, an even more dynamic reduction of the signal bandwidth is achieved. Figure 9.4 presents the bit stream format for the MPEG layer II system.

The MPEG from ISO has also defined the layer III and layer IV systems. As Section 9.2.2.3 discusses, DVB has decided to support the layer I and layer II systems only. Therefore, it is outside the scope of this section to describe the other systems as well.

9.2.2.2 Audio decoding

The MPEG audio decoding system more or less reverses the encoding procedure. The audio decoding systems for layer I and layer II are basically built up the same way (see Figure 9.5). First, the error correction information is decoded. Hence, detected bit errors are corrected. By means of the decoded values of the scale factors together with the decoded value of the allocated number of bits per sample, the subbands can be reconstructed. Next, an inverse filter bank assembles these subbands into the original digital signal. Finally, a D/A converter is needed to provide the required analog signal to the loudspeaker(s).

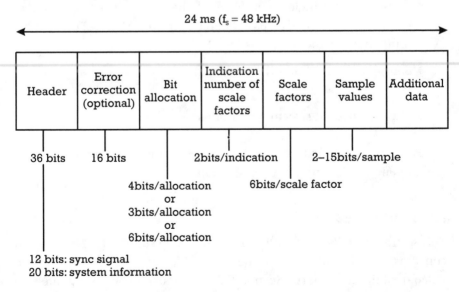

Figure 9.4 Bit stream format of the MPEG layer II system (mono).

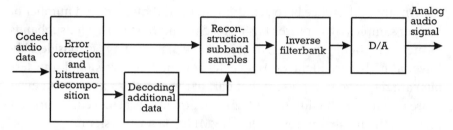

Figure 9.5 Functional description of the MPEG layer I and layer II audio decoding systems (mono).

9.2.2.3 DVB guidelines for audio coding

Because the MPEG of ISO had already defined a sufficient standard for the digital coding of audio signals and because this standard already was and still is applied in numerous applications worldwide, DVB has decided to adopt the MPEG standard rather than specify a new system. As mentioned in Section 9.1, DVB has produced guidelines for the implementation of the MPEG-2 digital audio coding system. The mandatory guidelines are summarized as follows:

▶ MPEG-2 layer I and layer II are supported by the IRD;

▶ The use of layer II is recommended for the encoded bit stream;

▶ IRDs support single-channel, dual-channel, joint-stereo, stereo, and the extraction of at least a stereo pair from MPEG-2 compatible multichannel audio;

▶ Sampling rates of 32 kHz, 44.1 kHz, and 48 kHz are supported by IRDs;

▶ The encoded bit stream does not use emphasis.

For a more detailed explanation of the (mandatory) guidelines, see the implementation guidelines document.

9.2.3 Video coding

This section describes the MPEG-2 video coding standard. This standard concerns the encoding and decoding of video signals and the several video qualities it supports. Section 9.2.3.4 discusses the DVB guidelines for the application of this standard.

9.2.3.1 Video encoding

The DVB guidelines for encoding video signals conform to the ISO/IEC IS 13818-2 standard [3]. This standard describes MPEG-2 video coding, which is applied typically in television studios and broadcasting. MPEG-2 makes use of redundancy reduction. The main reasons for DVB to adopt the work of MPEG are that it supports several video qualities up to HDTV programs and that it features high flexibility. Figure 9.6 presents a functional description of the MPEG-2 encoder.

In principle, the encoding system reduces the bandwidth by subtracting successive parts of the digital video signal. In case these successive parts are equal (e.g., the television screen is only showing one color), no information is encoded. At the decoder the subtracted information is added to the next part of the signal again to reconstruct the complete digital video signal. The MPEG-2 standard prescribes a subtraction taking place at the same frequency as that at which a picture on the television screen is updated.

At the input of the encoder, successive groups of digital video information are reordered in order to benefit from the equality of successive groups of information. This process is followed by a *discrete cosine*

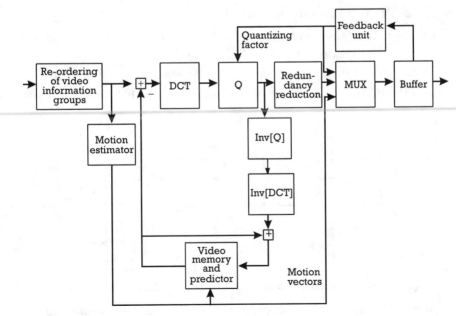

Figure 9.6 Functional description of the MPEG-2 video encoding system.

transformation (DCT). The DCT not only maps the video signal into the frequency domain, but the division of the amplitudes in the frequency domain shows less correlation as well. Next, the signal is quantized (eight bits), after which redundancy reduction by means of a Huffman code is applied.

The feedback circuit includes a video memory and a predictor. The video memory introduces a time delay to enable the subtraction of successive parts of the signal with a fixed length. The predictor is supported by a motion estimator, which produces motion vectors as an output. A motion vector indicates in which direction an object displayed on the television screen is moving.

A buffer is used to ensure a constant bit rate at the output of the decoder. In case the buffer suffers from an overload, the number of quantizing levels is decreased. Hence, the buffer will be provided with less data. The feedback from the buffer to the quantizer is represented by a quantizing factor, which allows an efficient use of the buffer. By providing the motion vectors, the output of the redundancy reduction, and the quantizing factors to a multiplexer, the decoder is able to reverse the encoding process.

9.2.3.2 Video decoding

The decoder basically reverses the encoding process. Figure 9.7 presents a functional description of the MPEG-2 decoder.

The encoded digital video signal with constant bit rate is to provide a buffer. The bit stream at the output of the buffer has the required variable

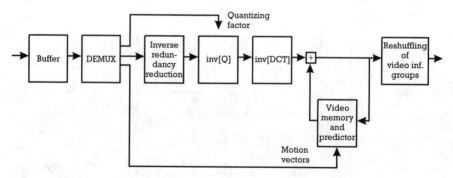

Figure 9.7 Functional description of the MPEG-2 video decoding system.

bit rate, which, on its turn, is provided to a demultiplexer. The demultiplexer separates the information concerning the quantizing factors and motion vectors from the actual video information. Next, in the process of redundancy reduction, the quantizing process (by means of the appropriate quantizing factors) and the DCT are reversed. The following step is the addition of the predicted video signal to the current video signal. The prediction process is supported by the motion vectors, which were included at the encoder. Finally, the groups of video information are reshuffled in the right order again.

By including a quantizer in the coding system, quantizing errors and thus a higher bit error rate may occur. This can result in a noisy picture on the television screen. In case the transmission of the digital video signal suffers from interference as well (e.g., in satellite or terrestrial broadcasting), no picture may be displayed on the television screen at all. To make the coding system more flexible, a *signal-to-noise ratio* (SNR) scalability is added to the system. SNR scalability introduces the ability to separate high-priority and low-priority bit streams. In case the bit error rate exceeds a certain threshold, only the high priority bits are provided to the output. This results in a picture on the television screen with an acceptable level of noise.

MPEG-2 also provides spatial scalability. In this case it is possible to optimize the resolution of the picture on the television screen. This is achieved by processing a digital video signal with a high resolution, as well as with a basic resolution. The output of the decoder provides the signal with basic resolution when the bit error rate is too high. The use of spatial scalability enables the compatibility with HDTV systems.

9.2.3.3 Levels and profiles

MPEG-2 is a family of standards consisting of a number of combinations of "levels" and "profiles." For source coding four data formats (levels) varying from low-definition television (similar to the quality of home VCRs), standard-definition television (comparable with PAL, SECAM, and NTSC quality), enhanced-definition television (ITU-R BT.601), and HDTV are used. Different bit rates apply to each of these levels. The levels are listed as follows.

▶ The low level contains a quarter of the picture's input-format, which is defined by ITU-R Recommendation BT.601;

▶ Next, the main level complies with the input-format defined by ITU-R Recommendation BT.601;

▶ Next, the high-1440 level has a high-definition format with 1,440 samples/line;

▶ Finally, the high level has a high-definition format with 1,920 samples/line;

Additionally, MPEG-2 has defined five different profiles. Each profile contains its own set of compression tools, which all together form the actual coding system. The profiles are designed in such a way, that each profile adds several extra tools to the preceding profile. This implies that each profile contains more features than the preceding one and will cost more. The profiles are the following:

▶ The simple profile has the smallest number of tools;

▶ The main profile includes the simple profile's tools plus one (bidirectional prediction). This improves the quality at the same bit rate but requires more space for an IC device. A main profile-decoder can decode information at simple profile, as well as at main profile. This demonstrates the strength of the system;

▶ The two following profiles are the SNR scalable profile and the spatial scalable profile. By means of these two profiles a base layer and "top-up" signals can be distinguished. The top-up signals can either improve the SNR scalability or the resolution (spatial scalability), respectively;

The first four profiles sequentially encode the color difference signals;

▶ Finally, the high profile contains all the preceding profiles' tools plus the ability to simultaneously encode the color difference signals. This makes MPEG-2 a kind of "super system" designed for the most critical circumstances at which a high bit rate is not the limiting factor.

Because not all combinations of levels and profiles were considered to be necessary, 11 out of 20 combinations were selected by MPEG. These 11 combinations are referred to as the MPEG-2 conformance points. Table 9.1 presents an overview.

Table 9.1
MPEG-2 Levels and Profiles

Profiles	Low-Level	Main-Level	High-1440 Level	High-Level
Simple	—	720 × 576 (15 Mbps)	—	—
Main	352 × 288 (4 Mbps)	720 × 576 (15 Mbps)	1,440 × 1152 (60 Mbps)	1,920 × 1,152 (80 Mbps)
SNR scalable	352 × 288 (4 or 3 Mbps)	720 × 576 (15 or 10 Mbps)	—	—
Spatial scalable	—	—	1,440 × 1,152 or 720 × 576 (60 or 40.15 Mbps)	—
High	—	720 × 576 or 352 × 288 (20 or 15.40 Mbps)	1,440 × 1,152 or 720 × 576 (80 or 60.20 Mbps)	1,920 × 1,152 or 960 × 576 (100 or 80.25 Mbps)

9.2.3.4   DVB guidelines for video coding

DVB has produced guidelines (ETR 154) for the use of the MPEG-2 digital video coding system. A summary of the mandatory guidelines is presented as follows:

- MPEG-2 main profile at main level (MP@ML) is used;

- The frame rate is 25 Hz;

- Encoded pictures may have either 4:3, 16:9, or 2.21:1 aspect ratios;

- IRDs (*integrated receiver decoder*) support 4:3 and 16:9 and optionally 2.21:1 aspect ratios;

- IRDs support the use of pan vectors to allow a 4:3 monitor to give a full-screen display of a 16:9 coded picture;

- IRDs support a full-screen display of 720 × 576 pixels (and a nominal full-screen display of 704 × 576 pixels);

- IRDs provide appropriate up-conversion to produce a full-screen display of 544 × 576 pixels and 480 × 576 pixels and a nominal full-screen display of 352 × 288 pixels.

For a more detailed explanation of the (mandatory) guidelines, see the implementation guidelines document.

ISO has also produced specifications for MPEG-3 and MPEG-4 digital coding systems. As DVB has decided to use MPEG-2, it is outside the scope of this section to describe the other systems as well.

9.2.4 Systems

The MPEG coded audio and video signals have to be transmitted in a comprehensive way. The MPEG-2 systems standard provides a multiplex and the addition of the required synchronization information. This section describes the MPEG-2 systems standard, as well as the DVB guidelines for its application.

9.2.4.1 Multiplex

Television programs technically consist of three elements: audio and video information as well as additional information to support these programs. These elements have to be provided to the IRD in an orderly way. For this purpose MPEG-2 has defined a standard [4], referred to as the MPEG-2 systems. This standard describes the multiplexing of the (MPEG) encoded audio signal, the (MPEG) encoded video signal, and the required additional information. Figure 9.8 presents a functional representation of the MPEG-2 systems.

The encoded audio signal is provided to a packetizer, which produces a stream of standardized packets, each including a header, an additional header (optional), and encoded audio information. This stream is called the *packetized elementary stream* (PES). As it concerns an audio signal, this PES is referred to as the audio PES. The same process applies to the encoded video signal and the additional data. Next, these three PESs are provided to a multiplexer. The multiplexer eventually produces a standardized data stream, including a header, an adaptation field (optional), and a payload including the information from the several PESs. This stream is referred to as the *transport stream* (TS).

The MPEG-2 system also provides a multiplex to produce a *program stream* (PS). One of the differences between the TS and PS multiplexes is that the former allows the use of different time bases. Another difference is that the packets within the TS have a fixed length (188-byte), while a variable packet length is allowed in the PS. Finally, the TS is suitable for

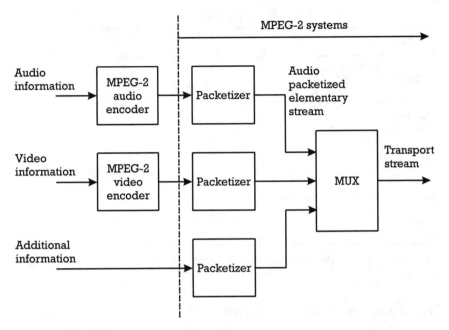

Figure 9.8 Functional representation of the MPEG-2 systems.

channels that suffer from a considerable level of interference (e.g., satellite channels). The PS is used for channels with low interference (e.g., storage of information on compact disks). For this reason it is preferable to use the TS for satellite, cable, and terrestrial channels, rather than the PS.

9.2.4.2 Synchronization

End-to-end synchronization of audio, video, and additional information is needed. Not only do the individual audio and video signals need to be synchronized, but both video signals and associated audio have to be synchronized as well. For example, the voices of the actors have to comply with the picture on the television screen. As mentioned in Section 9.2.4.1, two types of information streams are used: the TS and the PES. The data formats of both streams are presented in more detail in Figure 9.9.

Each TS packet's header contains a sync byte and a packet ID. The latter indicates what kind of program the payload contains (e.g., a pay-TV program). Optionally, synchronization of video signals and associated audio (i.e., the synchronization of the video PESs and the audio PESs) can

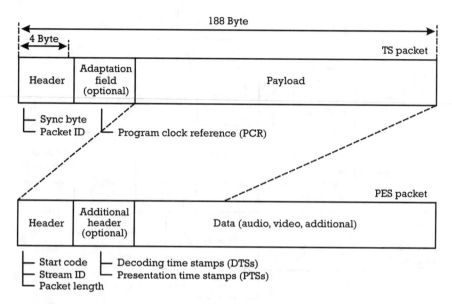

Figure 9.9   Data format of the TS and the PES.

be achieved by including an adaptation field in the TS packet. The adaptation field contains, among other things, a *program clock reference* (PCR). At the receiving end, the PCR is extracted from the TS and compared with the *system time clock* (STC) by means of a feedback circuit. This allows the regeneration of the clock signal, which is used for the synchronization of the decoding process.

Each audio and video PES packet contains a header, including bits concerning the start code, the stream ID, and the packet length. Optionally, a PES packet can contain information for the synchronization of the individual audio and video signals. For this purpose a *decoding time stamp* (DTS) and a *presentation time stamp* (PTS) can be included in an additional header.

The PTS is needed to ensure that the audio and video signals are provided to the loudspeaker(s) and the television screen, respectively, at the right time. The DTS indicates when the received data has to be loaded from a buffer into the audio or video decoder.

9.2.4.3   DVB guidelines for systems

DVB has produced guidelines (ETR 154) for the use of the MPEG-2 systems. A summary of the mandatory guidelines is presented as follows.

▶ MPEG-2 TS is used;

▶ *Service information* (SI) is based on MPEG-2 *program-specific informa-tion* (PSI), see Section 9.3;

▶ Scrambling is as defined in ETR 289;

▶ CA uses the MPEG-2 CA CA_descriptor;

▶ Partial transport streams are used for digital VCR applications.

For a more detailed explanation of the (mandatory) guidelines, see the implementations guidelines document.

9.3 DVB service information

This section discusses the SI that is included in the MPEG-2 standard as well as the mandatory and optional service information specified by DVB.

9.3.1 Service information

The MPEG-2 systems standard specifies SI data to enable automatic con-figuration of the IRD to demultiplex and decode the various streams of programs within the multiplex. This data is referred to by MPEG-2 as PSI. DVB has specified additional SI (ETS 300 468 [5]) to complement the PSI by providing data to aid automatic tuning of IRDs and additional data intended for display to the user.

SI data, which forms a part of DVB bit streams, provides the user with information to assist in selection of services and/or events. An event is a grouping of elementary broadcast data streams with a defined start and end time belonging to a common service, such as the first quarter of a bas-ketball game or a commercial. The specifications define that the IRD can automatically configure itself for the selected service. Moreover, SI can be used for VCR applications. As DVB did not specify the manner of present-ing information on the television screen, manufacturers in this case have a freedom of choice in presentation methods.

A recent development in DTV is the introduction of *electronic program guides* (EPGs). The definition of an EPG was considered outside DVB's scope. However, the specified data contained within the SI may be used as a basis for an EPG. This introduces another freedom of choice to the marketplace.

As stated earlier, the MPEG-2 systems standard specifies PSI. Both MPEG-2-defined PSI and DVB-defined SI (mandatory and optional) are discussed as SI in their entirety.

9.3.2   MPEG-2 defined service information

The MPEG-2 PSI data is structured as four distinct tables. The first table, called the *program association table* (PAT), indicates the packet ID values of the TS packets for each service in the multiplex. These values indicate the location of the corresponding *program map table* (PMT). The PMT identifies and indicates the locations of the streams that make up each service and the location of the PCR fields for a service. Additionally, the PMT gives the location of the *network information table* (NIT). The location of the NIT is defined by DVB in compliance with the MPEG-2 systems standard. For this reason, this NIT actually is not considered part of the MPEG-2 PSI data.

Finally, the *CA table* (CAT) provides the information on the CAMS used in the multiplex. As DVB has decided not to standardize the CAMS, different CAMSs may be used. Moreover, this information is private. For these reasons, the CAT is not specified. However, the CAT at least includes the location of the *entitlement management messages* (EMMs) stream, when applicable. The EMMs are private CA data that specify the authorization levels or the services of specific IRDs. They may be addressed to an individual IRD or to groups of IRDs.

9.3.3   DVB-defined service information (mandatory)

In addition to the NIT mentioned in Section 9.3.2, DVB has specified seven more tables, of which three are mandatory and four are optional [6]. The SI is needed to provide the identification of services and events for the user. The PSI (*program specific information*) (PAT, PMT, and CAT) only gives information concerning the multiplex in which it is contained. The additional SI can be used to provide information on services and events carried by different multiplexes, as well as on other networks.

The first additional table is defined as the *service description table* (SDT) and contains data describing, for example, the names and provider of services in the system. Next, the *event information table* (EIT) contains data related to events or programs such as the duration, the start time, or the event name. Moreover, the EIT allows the transmission of different kinds of event information. The table, referred to as the *time and date table* (TDT), gives information relating to the present time and date. Because of the frequent updating of this information, the TDT is defined as a separate table.

9.3.4 DVB-defined service information (optional)

The NIT of other delivery systems and the SDT and EIT of other TSs can be defined but are not really considered optional. The first option is formed by the *bouquet association table* (BAT). The term *bouquet* is used to describe a collection of services marketed as a single entity. The BAT provides a list of services for each bouquet as well as the bouquet's name. Second, the *running status table* (RST) provides data concerning the (running or not running) status of an event. Additionally, it updates this information and allows timely automatic switching to events. Next, the *stuffing table* (ST) is used to invalidate existing sections. A section is a syntactic structure, which is used for mapping all MPEG-2 PSI tables and DVB defined SI tables into TS Packets. Finally, the *time offset table* (TOT) provides information concerning the present time and data and local time offset. Due to the frequent updating of the time information, the TOT is defined as a separate table. Table 9.2 presents an overview of all SI tables.

Table 9.2
Service Information

	MPEG-2 PSI	DVB SI (Mandatory)	DVB SI (Optional)
Network information	PAT	NIT	NIT*
Bouquet information	CAT	—	BAT
Service description	PMT	SDT	SDT**
Event information	—	EIT	EIT**
Running status	—	TDT	RST
Stuffing	—	—	ST
Time offset	—	—	TOT

* Other delivery system.

** Other TS.

For the allocation of the SI codes for DVB systems, the technical specifications in the standard document, as well as the concerning requirements document, [7] can be consulted.

9.4   DVB teletext

This section discusses the background and basic elements of teletext and explains the DVB teletext system itself.

9.4.1   Elements of teletext

The first teletext systems were introduced in the United Kingdom by the BBC (Ceefax) and IBA-ITV (Oracle) in the 1970s [8]. The first agreed technical specification appeared in 1974, after which several systems with different specifications were developed. A teletext system is used as a value-added service, carrying extra information via television transmission systems. The teletext information is accommodated within the television signal. In fact, it is included in that part of the video information that is not visibly displayed on the television screen. Hence, no additional bandwidth is required.

A television set including a teletext decoder is capable of reconstructing written information and displaying it on the screen. The teletext information is often presented as a page. The teletext system supports a maximum of 800 single pages, of which not all may be used at any one time. The user can access a page by just selecting a three-figure number on a control panel. Next, the information related to the selected page is extracted from the incoming video information flow. After this short interval, the teletext information is presented on the screen. A rich amount of information services can be supported (e.g., weather forecast, travel information, stock exchange, and electronic newspaper). Moreover, these systems can supplement normal television programs with linked pages or even subtitles.

9.4.2   DVB teletext system

The ITU has specified a standard [9] for teletext systems. This standard, also known as EBU teletext [10], applies to analog teletext systems. The DVB standard (ETS 300 472 [11]) for a digital teletext system specifies the method by which these analog systems may be carried in DVB bit streams. Therefore, this transport mechanism, among others, has to support the transcoding of the teletext data into the *vertical blanking interval* (VBI) of analog video. Moreover, the transcoded signal must be compatible with existing television receivers with teletext decoders. A comparison can be made to the development of black-and-white television into color television. For color video coding black-and-white television sets were required to decode color video signals into black-and-white pictures on the screen.

The teletext data is conveyed in the PES packets. The PES packets in turn are carried by the MPEG-2 TS packets. The PMT for a specific service contains the packet ID of a teletext stream associated with that service. A service is allowed to include more than one teletext data stream. In this case, the SI contains information to distinguish both streams.

The DVB teletext standard does neither specify how a teletext decoder should be implemented, nor does it preclude any other architecture. It only specifies a conceptual model for decoding, which the bit stream is required to satisfy. This allows another freedom of choice for the market. The decoder model describes a teletext access unit, which is defined as a teletext data packet. A PTS defines the time at which the decoded text is intended to appear on the screen. In case of transcoding it indicates the time at which the access unit needs to be inserted in the VBI.

The decoder incorporates two linked teletext buffers. For a direct coding process, access units are extracted from the last buffer instantaneously as soon as a complete access unit is available. In the case of a transcoding process, this is the case whenever an appropriate video line is available in the associated video. Both processes require that the system time clock has reached the value of the PTS associated with this or any previous access unit. It has to be stated that if the transcoding model is obeyed, the direct decoding process is always satisfied.

9.5 DVB subtitling system

DVB has developed a specific system for the representation of graphical objects. This system is referred to as the DVB subtitling system. Screen definition and color coding are elements of this system. This section explains these elements and the DVB subtitling system itself.

9.5.1 Elements of DVB subtitling

To represent a graphical object on a television screen, this object's position on the screen as well as its color coding needs to be defined. This subsection describes the DVB specifications for these basic elements.

9.5.1.1 Screen definition

The representation of a graphical object (e.g., a logo, map, or subtitle) on the television screen is constructed by means of an encoded string of data bytes referred to as pixel-data. Each object has its own unique ID number. Individual graphical objects can be put on the screen at independent positions, so that flexibility is introduced. Because various screen layouts can share the same objects, efficiency is achieved as well.

A display is built up of regions. These are rectangular areas on the screen, with specified positions, in which objects are shown. Alternative screen layouts, defined as different page compositions, may use the same region (or any other graphical elements) without the need to convey that region for each screen layout in a separate way. This is useful when using, for example, the same logo on the screen in case subtitles in several languages are provided. This process is supported by the use of an (optional) ancillary page, which carries the elements shared by different screen layouts. The page composition of a screen layout, in its turn, is carried by the composition page.

Moreover, a shared region may be shown at different locations on different screen layouts. The position at which a region is shown on the screen is defined in the page composition. Several page compositions may be carried simultaneously in the bit stream, but only one page composition can be active at a time. Using the same example, this implies that only one language (plus logo) can be selected at a time.

9.5.1.2 Color coding

In order to translate the objects' pseudo colors into the correct colors on the screen, in each region a *color look-up table* (CLUT) is applied. In the rare case that one CLUT is not sufficient to process this translation, the objects can be horizontally split into smaller objects. The smaller objects, combined in separate regions, now require no more than one CLUT per region. The possible translations are defined by a family of CLUTs, listed in Table 9.3.

Table 9.3
CLUT Family

Entries	Map Tables
One CLUT with four entries (2-bit/entry) (mandatory)	—
One CLUT with 16 entries (4-bit/entry) (optional)	A map-table that assigns four entries of the 16-entry CLUT to pixel-data that uses a 2-bit per pixel coding scheme
One CLUT with 256 entries (8-bit/entry) (optional)	A map-table that assigns four entries of the 256-entry CLUT to pixel-data that uses a 2-bit per pixel coding scheme
	A map table that assigns 16 entries of the 256-entry CLUT to pixel-data that uses a 4-bit per pixel coding scheme

For the DVB subtitling system the four-entry CLUT is mandatory, while the CLUTs with 16 and 256 entries may be supported. For graphics that are basically monochromous (e.g., subtitles), a palette of four colors (which corresponds with the four-entry CLUT) should be sufficient. In case a more colorful picture (e.g., a cartoon movie) needs to be supported, several four-entry CLUTs can be combined. For this purpose, each graphical unit can be divided into several regions, corresponding with the total number of four-entry CLUTs. Next, a different color scheme in each of the regions can be applied. Alternatively, the colors in the entries can be redefined. This allows, for example, a red-blue-green scheme to be changed into a black-blue-yellow scheme. A combination of both options just described, is possible as well.

The coding efficiency can be increased by the application of map tables. Consecutive strings of, for example, 4-bit coded pixels that use a limited number of colors (e.g., four colors) can be represented by a 2-bit code. In case a 16-entry CLUT is used, the map table informs the decoder which entries of this 16-entry CLUT are to be used. This implies that the 2-bit codes are mapped on a 4-bit/entry CLUT. If the number of colors remains the same, but different colors are used, in that part of the pixel-data the coding may switch to a 2-bit/pixel using another map table. The final coding efficiency depends on the number of pixels that can be coded without changing the code mode or map table.

9.5.2 DVB subtitling system

DVB has constructed a model (prETS 300 743 [12]) for the process-ing required for the interpretation of subtitling streams. This model defines the constraints for the verification of the validity of these streams. Figure 9.10 presents a typical implementation of a DVB subti-tling decoder.

The MPEG-2 TS packets are provided to the input of the decoding process. A packet ID filter is used to select the subtitle-related TS packets. Next, these packets enter into a transport buffer. The size of this buffer is 512 bytes. In case any data has entered the buffer, data is removed at a data rate of 192 Kbps. This data stream is provided to the actual subtitle decoder. As such, the decoder includes a preprocessor and filters, a coded data buffer, a subtitle processor, and a composition buffer.

The preprocessor strips off the TS packet headers and the proper PES packet headers. The PES header is mainly used to accommodate a PTS for the subtitling data. This PTS is passed on to the next stages of the process. The PES packets encapsulate page segments. It is possible that page seg-ments exceed the capacity of a PES packet. In this case, segments for one display time are split over several PES packets. Each of these packets incorporate the same PTS value. A filter is used to provide the segments, related to one page, to the next stage. Next, these selected segments are provided to a Coded data buffer with a size of 24 kByte. The (complete) segments are removed and decoded in an instantaneous process. The removal of segments out of the coded data buffer stops when a segment produces pixel-data. No segments are removed until all pixels have been loaded (at a rate of 512 Kbps) into the pixel display buffer.

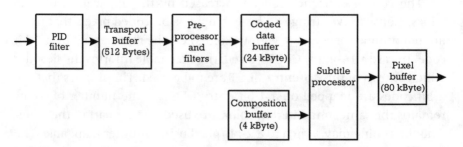

Figure 9.10   DVB subtitle decoder model.

The pixel display buffer has a capacity of 80 kByte of which 60 kByte can be assigned to pixels that are displayed on the screen simultaneously. The remaining capacity can be used to hold pixel-data for future display. The composition buffer contains all display data structures other than the displayed graphical objects. This concerns information on page composition, region composition, and CLUT definition. DVB has not specified the control of the various buffers by the encoder. This is left to the market.

Another constraint is defined in case a real-time subtitling decoder is applied. This type of decoder allows the immediate transfer of decoded data to the display. This is achieved by storing the coded data in a buffer and continuously decoding this data and generating the pixel values in real time. This requires a (larger) coded data buffer with a capacity of 48 kByte.

9.6   Summary and conclusions

The MPEG-2 standard provides a "tool kit" of compression and transmission techniques. For any application, users must choose which tool to use. This chapter has discussed how DVB implements the MPEG-2 standard for digital audio-visual coding by providing guidelines. In addition to this standard, DVB has specified extra SI to assist the user in selecting special services. Beside these coding techniques, DVB has produced specifications for additional services, such as teletext and subtitling services. The DVB digital teletext system is compatible with analog teletext systems. A great variety of services for displaying graphical objects (e.g., logos and subtitles) on the television screen is introduced by the DVB subtitling system. Table 9.4 presents an overview of source coding and additional services.

Moreover, this chapter explains for each specification (where relevant) the freedoms of choice for the marketplace. This should give service providers and/or manufacturers an incentive to distinguish themselves. The user could benefit from this, because a variety of services with distinct quality could be placed at his or her disposal.

Table 9.4
Source Coding and Additional Services

Standard	Application	Features
MPEG-2 audio	Audio Coding (layer I and II)	Single-channel, dual-channel, (joint) stereo, at least one stereo pair of multilingual sound, and surround sound (optional)
MPEG-2 video	Video coding (MP@ML)	SNR scalability and spatial scalability
MPEG-2 systems	SI (PSI)	Network, bouquet, and SI
DVB SI	SI	Mandatory: network, service, and event information and running status
		Optional: stuffing and time offset
DVB teletext	Analog TT in DVB bit stream	No special features
DVB subtitling	Graphical object coding	Screen definition and color coding

References

[1] EBU/CENELEC/ETSI JTC, *Digital Video Broadcasting (DVB); Implementation guidelines for the use of MPEG-2 systems, video and audio in satellite, cable and terrestrial broadcasting applications*, ETR 154, Second Edition, October, 1996.

[2] ISO/IEC, *Coding of moving pictures and associated audio —Part 3: Audio*, IS 13818-3.

[3] ISO/IEC, *Coding of moving pictures and associated audio—Part 2: Video*, IS 13818-2, 1994.

[4] ISO/IEC, *Coding of moving pictures and associated audio—Part 1: Systems*, IS 13818-1, 1994.

[5] EBU/CENELEC/ETSI-JTC, *Digital Video Broadcasting (DVB); Specification for Service Information (SI) in DVB Systems*, ETS 300 468, Second Edition, January, 1997.

[6] EBU/ETSI-JTC, Digital Video Broadcasting (DVB); *Guidelines on Implementation and Usage of Service Information*, ETR 211, Final Draft, 5 February, 1997.

[7] EBU/ETSI-JTC, *Digital Video Broadcasting (DVB); Allocation of Service Information (SI) codes for Digital Video Broadcasting (DVB) systems*, ETR 162, October, 1995.

[8] Mazda, F.,*Telecommunications Engineers' Handbook*, 1993.

[9] ITU-R, *Recommendation 653: System B, 625/50 television systems*.

[10] EBU, *Teletext specification (625-line television systems)*, EBU SPB 492, 1992.

[11]   EBU/CENELEC/ETSI JTC, *Digital Video Broadcasting (DVB); Specification for conveying ITU-R System B Teletext in DVB bitstreams*, ETS 300 472, Second Edition, October, 1996.

[12]   EBU/CENELEC/ETSI JTC, *Digital Video Broadcasting (DVB); DVB Subtitling system*, DRAFT prETS 300 743, November, 1996.

Digital transmission

10.1 Introduction

The introduction of digital transmission technology enables a dramatic decrease in required bandwidth per channel. In the context of television services this implies that more television channels will be available to provide the home user with programs and special services. Services such as pay-per-view and video-on-demand, for example, require a large number of television channels. Because of the increased number of available channels it is expected that the costs per channel will drop. However, this depends on the investments that have to be made to enable the actual implementation of the digital transmission technology. Standards can play an important role in the establishment of economies of scale in order to obtain a return on investment.

This chapter discusses the DVB specifications for digital transmission technology, which were standardized by ETSI. DVB has specified several systems for digital communications via satellite, CATV, and terrestrial networks. The basic elements and the several (sub)systems pass the revue per communication system. Moreover, the channel encoding and decoding processes are explained at a functional level. The presence of *additive white gaussian noise* (AWGN) is assumed for each channel, unless stated otherwise. Information concerning the system performance can be obtained from the ETSI standards papers. Consult [1] for a more detailed study of the main DVB digital transmission systems.

10.2   DVB satellite

Television signals can be provided via various transmission networks. This section discusses the DVB specifications for satellite communications, first explaining the basic elements of satellite communication and then describing the DVB channel encoding and decoding systems.

10.2.1   Elements of satellite communications

Several elements play a role in satellite communications. The several elements discussed in this section concern typical transmission characteristics, which have to be regarded when specifying a satellite communications system.

10.2.1.1   Transmission medium

By providing point-to-multipoint communications from a point in space, satellites typically have the ability to create simultaneous links to users on Earth. Moreover, with satellites, capacity can be dynamically allocated in correspondence to the users' needs.

Satellites that are used for broadcasting purposes are located in a *geostationary orbit* near the equator at an altitude of about 36,000 km. Because these satellites move as fast and in the same direction as the Earth rotates, perceived from the Earth's surface, they seem to hang still at a fixed point.

The energy needed for transmission is supplied by the satellite's solar system. As a result of the solar system's low efficiency, the power of the output signal is limited. However, the rich availability of bandwidth counters this limitation. Today's satellite systems use channel bandwidths from 26 MHz to 54 MHz.

The ITU, which acts under the authority of the United Nations (UN), has developed rules and guidelines called *Radio Regulations* [2]. Since 1903 a series of international radio conferences have been held. The most recent was the 1995 World Radio Conference (WRC-95). Table 10.1 lists the frequencies allocated to *broadcasting satellite services* (BSS), as well as the corresponding geographical areas.

10.2.1.2 Satellite uplink/downlink

The baseband signal is processed and transmitted to the satellite by a modulated *radio frequency* (RF) carrier. The RF carrier is transmitted from the Earth station to the satellite via the *uplink* frequency spectrum (Figure 10.1). Next, the RF carrier is sent back to Earth via the *downlink* frequency spectrum.

In order to avoid interference, the uplink and downlink are operated on different frequencies. Beside using BSS frequencies, sometimes frequencies allocated to FSSs are used for the uplink. A BSS, for example, could use a 14.0–14.5 GHz uplink (FSS) and an 11.7–12.5 GHz downlink (BSS). The function of the satellite itself can be thought of as a large repeater in space. It simply receives an RF signal, amplifies the RF signal, translates the signal frequency, and sends the signal back to Earth.

Table 10.1
Frequency Allocations to Broadcasting Satellite Services Related to Television (Downlink)

Frequency Range (GHz)	Restriction
2.52–2.655	c
11.7–12.2	1, 3 only
12.2–12.5	1, 2 only
12.5–12.7	2, 3c only
12.7–12.75	3c only
21.4–22	1, 3 only
40.5–42.5	
84–86	

Notes:
c = community reception only;
1 = (Region 1): Europe, Africa, former USSR, and Mongolia;
2 = (Region 2): North and South America and Greenland;
3 = (Region 3): Asia (except former USSR and Mongolia), Australia, and the Southwest Pacific.

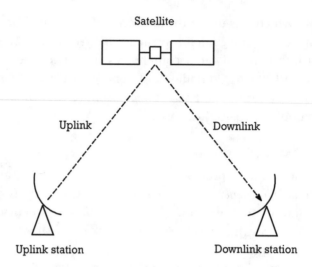

Figure 10.1   Satellite system.

10.2.1.3   Orthogonal polarization

The number of transponders can be increased by "reusing" frequencies. This is accomplished by means of orthogonal polarization, which allows two signals to be transmitted in the same frequency without interfering. The polarization of a radiated electromagnetic wave is the curve traced by the endpoint of the instantaneous electric field vector as observed along the direction of propagation. Polarization can be classified as linear, circular, or elliptical [3].

In case of linear polarization, the electric field vector oscillates along the horizontal or the (orthogonal) vertical line, respectively (Figure 10.2). As a result, the information capacity carried by the satellite can be doubled. EUTELSAT and ASTRA, among others, make use of this type of polarization. The electric field vector of a circularly modulated signal rotates clockwise or counterclockwise orthogonally in the direction of propagation. Hence, the electric field vector is constant in length but traces a circle. This type of polarization is applied in TV-SAT. In case of elliptical polarization, the electrical field vector describes a (counter)clockwise elliptical curve. At the receiving end the orthogonally polarized signals can be separated again.

In theory there is infinite isolation between orthogonal polarizations. Due to system imperfections and propagation influences, however, this is not the case in practical systems. Dual-polarized transmission requires a

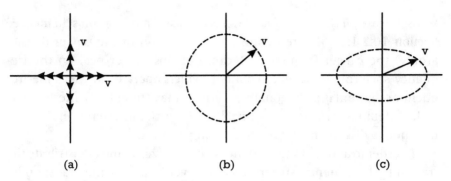

Figure 10.2   Types of orthogonal polarization: (a) linear, (b) circular, and (c) elliptical.

good level of isolation between two polarizations in order to keep the interference at an acceptable level. This can, for example, be achieved by employing an interleaved frequency plan as shown in Figure 10.3.

10.2.1.4   Energy dispersal

In general, the power density of a DTV signal is equally divided over its own bandwidth. It might occur, however, that for a period of time the bit stream of a television signal contains a sequence of either all ones or all

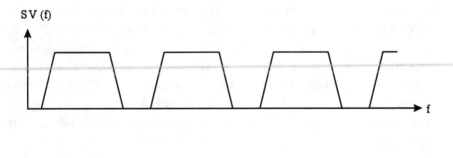

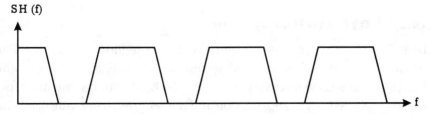

Figure 10.3   Frequency reuse by linear orthogonal polarization.

zeros. In case *phase shift keying* (PSK) is used as a modulation scheme (see Section 10.2.1.5), this results in a concentration of the power density around the carrier frequency. Because of this power peak in the frequency spectrum, a satellite channel, which operates in the same frequency range but is orthogonally polarized, may suffer from interference. If the satellite Earth station is unable to filter the interference, the user may notice disturbance of the television program.

In order to avoid a long sequence of ones or zeros and to distribute the transmitted frequency spectrum more evenly across the transponder bandwidth, the bit stream of the television signal is randomized. This means that the signal is more or less represented by alternating ones and zeros. A digital scrambling device changes the signal as it would concern a bit stream with a random structure. At the receiving end the signal is descrambled to obtain the original signal again.

10.2.1.5   Modulation

To transmit information over a bandpass channel, a baseband signal, which represents the information, is modulated on a carrier frequency. Digital information is assumed to be binary (1 and 0) and occurs at a rate of 1 bit per Tb seconds. Alternatively, the binary digits can be segmented into blocks consisting of m bits. Since there are $M = 2m$ blocks, M different signal values are required to represent the m-bit blocks unambiguously. Each m-bit block is called a symbol. The symbol duration is $T_s = mT_b$ seconds. This type of transmission is referred to as M-ary signaling.

The actual modulation can be achieved by varying the signal amplitude, frequency or phase. M-ary PSK is commonly used in digital satellite communications. The signal amplitude is constant and M different phases are used to represent M distinct symbol values. A constant amplitude is very important to the non- linear characteristic of the transponder. M-ary signaling allows transmission of m bit in a bandwidth of one Hertz.

10.2.2   DVB satellite systems

DVB has specified satellite systems for DTH satellite services, satellite services via cable television, and terrestrial broadcasting networks and SMATV. This section discusses the several DVB satellite systems at a functional level in which the application and the required bandwidths play an important role.

10.2.2.1 Direct-to-home system

Within DVB, satellite systems for digital multiprogram television and HDTV have been specified. One of these systems concerns the DTH system for consumer IRDs. The DTH system operates at 11/12 GHz and uses bandwidths in the range of 26 MHz to 54 MHz. Home users can receive the signal broadcast by the satellite directly, by means of a satellite dish (diameter 60cm). The specifications for this system have been standardized by ETSI (ETS 300 421 [4]).

10.2.2.2 Satellite services via cable television and terrestrial broadcasting networks

The same standard (ETS 400 421) applies to the transmission of satellite signals to a CATV head-end station. For various reasons, CATV network operators may want to decide for themselves which programs they want to provide to their users. At the cable head-end the signals are separated in a demultiplexer. Next, with the help of a remultiplexer the desired programs are compiled. Finally, the remultiplexed satellite signal is remodulated in order to be accommodated in an 8-MHz CATV channel to the home user IRD.

The same procedure also applies to terrestrial broadcasting networks. In this case, the satellite signal is remodulated in order to be provided to the home user IRD via, for example, an 8-MHz terrestrial channel.

10.2.2.3 Satellite master antenna television system

A SMATV system is defined as a system for the distribution of television and sound signals to households in one or more adjacent buildings [5]. These signals are received by a satellite receiving antenna and may be combined with terrestrial television signals. SMATV distribution systems are also known as community antenna installations or domestic television cable networks.

In correspondence to the CATV head-end, the desired satellite and terrestrial signals are demodulated, demultiplexed, remultiplexed, and remodulated according to the SMATV channel characteristics in the SMATV head-end. The DVB specifications (ETS 300 473 [6]) for the SMATV system offer two alternatives for providing signals to the home user IRD. The satellite signals can be distributed directly using frequency conversion (e.g., to the extended *intermediate frequency* (IF) band (0.95 GHz to 2.05 GHz)) or to the extended *super* (S)-band (0.23 GHz to

0.47 GHz)). The alternative is to first remodulate the satellite signal and distribute it via the SMATV network and second, use the same frequency conversion.

10.2.3 Channel encoding

This section explains how the constraints in the basic elements of satellite communication, which were discussed in Section 10.2.1, are respected by DVB in the specifications for channel encoding.

10.2.3.1 Encoding system

Before explaining the signal encoding process in more detail, a conceptual representation of the encoding system and modulation is provided in Figure 10.4.

The most important steps to adapt the TS to the satellite transmission medium are the following.

▶ Transport multiplex adaptation;

▶ Randomization for energy dispersal;

▶ Error correction coding and interleaving;

▶ Baseband shaping for modulation;

▶ Modulation.

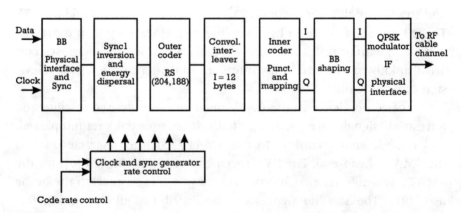

Figure 10.4 Conceptual encoding system description.

10.2.3.2 Transport multiplex adaptation

The satellite system is compatible with MPEG-2 coded signals [7]. This implicates that the modem transmission frame complies with the MPEG-2 multiplex transport packets. These packets consist of 188 bytes of which the first four bytes are used for the header. The header's first byte is reserved for the synchronization byte. The length of the packets, 188 bytes, was chosen to ensure compatibility with ATM transmission. ATM is considered to be an important future transmission technology and has already been introduced in some public broadcasting and telecommunications networks.

10.2.3.3 Energy dispersal scrambling

To avoid a concentration of the power density around the carrier frequency, the data of the MPEG-2 TS is randomized by a *pseudo random binary sequence* (PRBS) generator (Figure 10.5).

At the start of every eight transport packets, an initial sequence is loaded into the PRBS registers. Via an exclusive-or operation the first bit (i.e., MSB) at the output of the PRBS register is applied to the first bit of the first byte following the inverted MPEG-2 sync byte. The MPEG-2 sync bytes of the subsequent seven transport packets are not randomized in order to support other synchronization functions. Although the PBRS generation continues during this process, its output is disabled. The period of the PRBS sequence is 1,503 bytes.

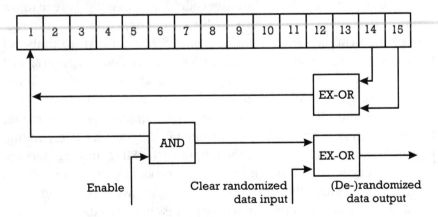

Figure 10.5 PRBS generator.

10.2.3.4 Inner coding

Digital transmission allows the use of *forward error correction* (FEC). The DVB satellite system requires a *quasi-error-free* (QEF) transmission. This means that less than one uncorrected error-event per hour is allowed. Hence, the *bit error ratio* (BER) must be within the range of $1*10^{-11}$ to $1*10^{-10}$ at the input of the MPEG-2 demultiplexer.

The satellite system incorporates two different error control procedures—an outer and inner coding. The latter is located closest to the satellite link. The outer coder uses a *Reed-Solomon* (RS) code. The 188 bytes packets are expanded with 16 redundant bits. Therefore this code is referred to as RS(204,188). The adding of these redundant bits allows up to eight erroneous bytes per packet. As a result, the BER may increase up to $2*10^{-4}$ at the input of the RS decoder in order to meet the required BER of $1*10^{-11}$ to $1*10^{-10}$ at the input of the MPEG-2 demultiplexer.

10.2.3.5 Convolutional interleaving

During satellite transmission lengthy burst errors for which the application of an error correction code is not sufficient may occur. By means of an interleaving process, adjacent symbols become separated. As a result mutilated packets are split up into individual errors. These errors can be corrected by the RS decoder at the receiving end. This procedure is referred to as convolutional interleaving. Figure 10.6 shows a conceptual representation of convolutional interleaving as applied by DVB.

The output packets from the outer coder are consecutively read into a *first-in, first-out* (FIFO) shift register, which contains M cells. The shift register is called a branch and the *interleaving depth* (I) refers to the number of branches the interleaver incorporates. DVB has specified $I = 12$ and $M = 17$ ($M = N/I$ and $N = 204$ bytes). As a result, adjacent mutilated bits in the channel are located at least 205 bytes apart from each other in the received TS after de-interleaving. In order to support synchronization, the (inverted) sync bytes are always routed in the branch corresponding to $I = 0$ of the interleaver. Next, the output of the FIFO shift registers are cyclically connected to the input of the inner coder by the output packet switch. This requires that the input and output switches are synchronized. At the receiving end, the whole process is reversed.

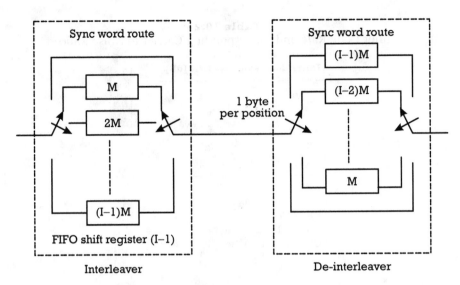

Figure 10.6 Convolutional interleaving.

10.2.3.6 Outer coding

A higher output power of a satellite signal has a beneficial effect on the BER. Because of technological and economical constraints, however, the satellite offers a medium power level. This is insufficient to achieve the required BER. To maintain the same BER, this implies that the Earth station satellite dish must have a larger diameter. However, especially in the case of DTH systems, the diameter of the satellite antennas must be small. A successful introduction of digital satellite technology requires low-cost home consumer antennas.

If the satellite dish diameter has a given value and the BER turns out to be high, the alternative to guarantee a QEF quality is to add error correction bits according to a convolutional code (Viterbi code) that doubles the total amount of bits. A more economical coding can be achieved by an additional process called puncturing. The redundancy of bits with respect to the useful information, which is referred to as the code rate, can now be chosen. For example, a code rate of 3/4 indicates that the total data contains 25% error correction bits and 75% useful data. Depending on the specific needs of satellite transmission, different code rates can be applied (see Table 10.2). In case the satellite produces a relatively high output

Table 10.2
Inner Code Rate and Corresponding Carrier to Noise Ratio

Inner Code Rate	C/N [dB]
1/2 np	4.1
2/3	5.8
3/4	6.8
5/6	7.8
7/8	8.4

Notes:
np = no puncturing;
B = 33 Mhz;
BER = $2*10^{-4}$ after Viterbi;
QEF (BER = $1*10^{-11}$ to $1*10^{-10}$) after RS.

signal power, the number of redundant bits can be kept small. This allows maximum error protection efficiency and a flexible implementation of the DVB satellite specifications.

10.2.3.7 Filtering

Prior to modulation, the digital signal is filtered so that it does not exceed the satellite channel's bandwidth. Exceeding this bandwidth could lead to interference with adjacent channels. According to the Nyquist (pulse shaping) criterion, the bandwidth (B) occupied by the pulse spectrum is $B = (r_s/2)(1+alpha)$, in which r_s represents the symbol rate, and alpha is the filters' roll-off factor, where 0<alpha<1. Theoretically, a channel bandwidth of at least $r_s/2$ is required to accommodate the signal. In practice, however, the signal is formed by a raised cosine, which implies that the signal bandwidth is larger than $r_s/2$. Hence, a guard interval between two adjacent channels is required. If the guard interval is sufficiently large, a raised cosine filter can be used (Figure 10.7). This results in an acceptable level of interference.

DVB has specified a square root raised cosine filter with roll-off factor alpha = 0.35.

10.2.3.8 Modulation

The DVB digital satellite system uses *quadrature PSK* (QPSK) where the amplitude has four phase states (M = 4), and together these phases can

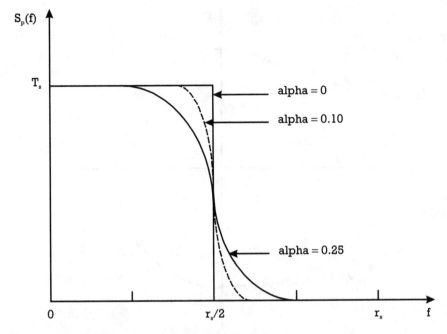

Figure 10.7 Raised cosine spectrum.

carry information that is represented by two bits ($m = 2$). This implies transmission of up to 2 bits in a bandwidth of one hertz. The actual transmission efficiency depends on the error coding applied. Figure 10.8 presents the QPSK constellation diagram. The Gray coding used defines that a phase shift of ±90 degrees implies that the digital representation of the phase changes one bit only.

10.2.4 Channel decoding

Section 10.2.3 explained the DVB encoding system for satellite communications. At the receiving end the signal needs to be decoded again in order to obtain the original signal. Hence, this section discusses the DVB specifications for the channel decoding.

10.2.4.1 Decoding system

At the receiving end, with the help of the recovered carrier and clock signals and sync signal, the decoding system more or less reverses the coding

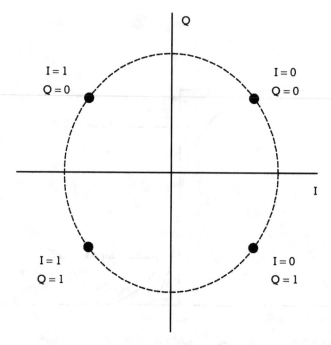

Figure 10.8   QPSK constellation diagram.

process. Hence, the decoding system (Figure 10.9) incorporates the following:

▶ Demodulation;

▶ Baseband reshaping and carrier and clock recovery;

▶ Inner error correction decoding;

▶ Synchronization decoding;

▶ Outer error correction decoding and de-interleaving;

▶ Derandomization for energy dispersal;

▶ Transport multiplex adaptation.

10.2.4.2   Demodulator

At the input of the receiving end the QPSK demodulator detects the phase of the carrier signal after which the symbol information can be

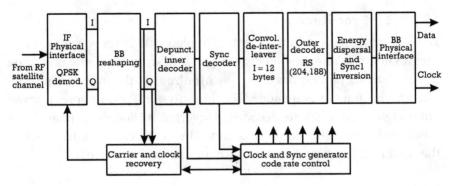

Figure 10.9 Conceptual decoding system description.

demodulated. Because the carrier signal can have four different phases (each with a 90-degree difference), a selection procedure is used to detect the correct phase in a maximum of two steps. The first step detects a phase error of ±90 degrees. In the next step a possibly remaining 180-degree phase error can be detected. The actual phase error detection and correction is executed in the following decoding process.

10.2.4.3 Filtering and carrier and clock recovery

The demodulated digital pulses are reshaped by means of a complementary square root raised cosine filter. In compliance to the filter at the transmitting end, the roll-off factor alpha is 0.35. This results in an acceptable level of interference with adjacent satellite channels. The demodulator synchronization is achieved by means of a carrier and clock recovery unit, which makes use of a *phase-locked loop* (PLL). The PLL functions as a feed back circuit in order to lock on to the rhythm of the clock signal.

10.2.4.4 Viterbi decoder

The filtered signal is then provided to the inner decoder, which incorporates a Viterbi [8] decoder with flexible depuncturing of error correction bits. In a trial and error process the correct code rate and depuncturing for the decoding process is selected. Moreover, a ±90-degree phase error can be detected. Depending on the adopted code rate, a BER in the order of $1*10^{-2}$ to $1*10^{-1}$ at the input of the Viterbi decoder is allowed so as to obtain the required BER of $2*10^{-4}$ at the input of the RS decoder for QEF quality in the end.

10.2.4.5 Sync decoder

To reconstruct the data stream with complete 204 bytes packets for further RS demodulation and energy dispersal descrambling, the preceding de-interleaving process has to be synchronized. At the transmitting end, synchronization bits were added for this purpose. Furthermore, if seven out of eight sync pulses are decoded as inverted, a 180-degree phase error is detected. This error cannot be detected by the Viterbi decoder. Next, at the output of the sync decoder, the data stream is inverted.

10.2.4.6 De-interleaver and Reed Solomon decoder

The interleaving process is reversed at the receiving end by means of a de-interleaver. As described above, the de-interleaver is synchronized to regain the complete data packets. As mentioned earlier, the (de)-interleaving and RS (de)coding process enables the correction of burst errors. The BER at the input of the RS decoder must be $2*10^{-4}$ at the most in order to obtain a BER in the order of $1*10^{-11}$ to $1*10^{-10}$. This complies to the required QEF quality.

10.2.4.7 Energy dispersal descrambler

The energy dispersal descrambler finally recovers the MPEG-2 Transport Stream by reversing the scrambling procedure. The descrambler is initiated by the inverted sync byte of the first transport packet into a group of eight packets. Next, the TS is provided to an MPEG-2 demultiplexer, after which MPEG-2 source decoding follows. In the TS an additional bit is inserted directly after the sync byte. This bit is to indicate whether an error has occurred during transmission but has not been corrected during the error correction process.

10.3 DVB cable

Beside satellite communication systems, television signals can be provided via CATV networks. This section discusses the DVB specifications for cable communication, first explaining the basic elements of cable communication and then describing the DVB channel encoding and decoding systems.

10.3.1 Elements of cable communications

This section discusses the several elements that play a role in cable communications, including typical transmission characteristics, which have to be regarded when specifying a cable communication system.

10.3.1.1 Transmission medium

Within CATV networks digital video signals are typically transmitted via 8-MHz channels. The theoretical maximum symbol rate ($r_{s,max}$) depends on the roll-off factor (alpha) of the raised cosine filter. Hence, $r_{s,max} = 8\ \text{MHz}/(1 + \text{alpha})$. For alpha $= 0.15$ this implies $r_{s,max} = 6.96$ MBaud.

In cable networks the transmitted signals attenuate after traveling a certain distance through the network. In order to obtain an adequate S/N at the receiving end, the cable network is equipped with repeaters. A repeater filters the noise and amplifies the digital signals to the required power level for further transmission through the network.

10.3.1.2 Signal reflection

The reflection of signals is another typical aspect of cable communication. This occurs, for example, when cables are not ideally connected. As a result, the cable impedance is no longer characteristic. At the point of connection a part of the signal is reflected and travels back in the direction of its origin. The reflected signal may be reflected once again in the direction of the receiving end and is added to signals traveling in the same direction. Because of the attenuation of signals in general, the impact of these reflections at the receiving end is negligible.

10.3.1.3 Modulation

Satellite communication is subject to power limitations. These limitations do not apply to communication via CATV networks and therefore it is possible to modulate not only the phase, but the amplitude as well. For the DVB digital cable system an M-ary signaling (see Section 10.2.1.5) referred to as *quadrature amplitude modulation* (QAM) is applied. Hence, a larger number of bits/symbol is allowed. A modulation efficiency of 4 bits/symbol is achieved by 16-QAM, 5 bits/symbol is achieved by 32-QAM, and an efficiency of 6 bits/symbol is achieved by 64-QAM.

10.3.2 Channel encoding

This section describes how the constraints in the basic elements of cable communication are respected by DVB in the specifications for channel encoding.

10.3.2.1 Encoding system

The specifications for the DVB cable system (ETS 300 429 [9]) can be used transparently with the DVB satellite system (ETS 300 421). As described in Section 10.2.2.2, programs broadcast by satellite can be received in the CATV cable head-end. After channel adaptation, these programs can be provided to the home-user IRD. Hence, the conceptual representation shows much correspondence (see Figure 10.10). The common elements are presented in gray.

10.3.2.2 Byte to m-tuple conversion

Depending on the modulation efficiency of 2^m-QAM modulation, k bytes are mapped onto n symbols of m bits (m-tuple conversion), such that $8k = n*m$ because 1 byte consists of 8 bits. In case of 16-QAM, the modulation efficiency is 4 bits/symbol ($m = 4$) and two symbols ($k = 2$) of four bits each ($n = 4$) can be formed out of one byte. Before m-tuple conversion, the MPEG-2 transport packets contain 204 bytes. After the conversion, a transport packet contains 408 symbols (Figure 10.11). Correspondingly, for 32-QAM eight symbols can be formed out of every 5 bytes. For 64-QAM this results in four symbols out of every 3 bytes.

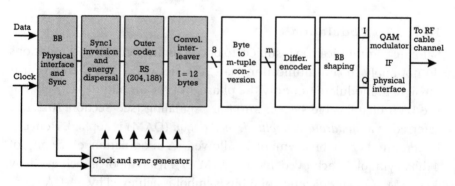

Figure 10.10 Conceptual encoding system description.

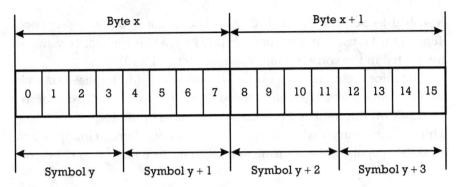

Figure 10.11   *m*-tuple conversion for 16-QAM.

10.3.2.3 Differential coding

After *m*-tuple conversion the symbols have to be prepared to be mapped in the QAM-constellation. This is achieved by the differential coding of the two *most significant bits* (MSBs) of each symbol. The MSBs define the quadrant in which the symbol is mapped (see Table 10.3).

10.3.2.4 Filtering and modulation

Before modulation, the digital signal is filtered. Corresponding to the DVB satellite system, a square root raised cosine filter is used. However, as a result of less available bandwidth per channel, for the DVB cable system a roll-off factor alpha = 0.15 is chosen.

In case of satellite communication, a maximum of four distinct symbol values can be distinguished in the QPSK(*quadrative phase shift keying*)-constellation diagram as result of QPSK modulation. These symbols all have the same frequency and amplitude, but have different phases. As

Table 10.3
MSBs Related to the Rotation in the QAM-Constellation Diagram

MSBs	Rotation	Quadrant
00	0°	1
01	−90°	4
10	+90°	2
11	180°	3

described in Section 10.3.1.3, CATV systems allow more bits/symbol. Hence, QAM is used. When, for example, 16-QAM is applied, 16 symbols are located in the constellation diagram (Figure 10.12).

The information in each of the 16 distinct amplitude/phase states is represented by 4 bits. This allows transmission of up to 4 bit/s in one hertz. For 32-QAM this implies 32 distinct states and a symbol length of 5 bits, which allows transmission of up to 5 bits/s in one hertz. Finally, when 64-QAM is applied, the transmission efficiency is 6 bits/s in one hertz.

10.3.3 Channel decoding

In Section 10.3.2, the DVB encoding system for cable communication is explained. To obtain the original signal again at the receiving end, the signal needs to be decoded. Hence, this section discusses the DVB specifications for the required channel decoding.

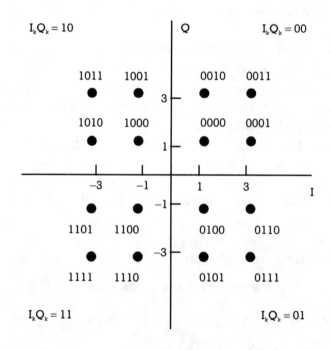

I_kQ_k are the two MSBs in each quadrant

Figure 10.12   16-QAM constellation diagram. (I_kQ_k are the two MSBs in each quadrant.)

10.3.3.1 Decoding system

The signal encoding procedure is reversed at the receiving end with the help of the recovered carrier and clock signals and sync signal. Figure 10.13 presents the conceptual decoding system description.

10.3.3.2 Demodulation and filtering

The QAM demodulator now must regain the distinct symbols. Hence, the correct phase/amplitude states of the symbols are detected. In correspondence to the satellite system, phase errors are corrected in the following process.

Next, the digital pulses of the input signal are reshaped by means of a complementary square root raised cosine filter with a roll-off factor alpha of 0.15. Hence, the interference with adjacent cable channels is restrained.

10.3.3.3 Carrier and clock recovery

The demodulation process is synchronized by means of a carrier and clock recovery unit. Corresponding to the DVB satellite system, a feedback circuit (PLL) is used to recover the carrier and clock signals. In contrast with the satellite system, the correction of ±90-degree and 180-degree phase errors is achieved by comparing the original carrier phase and the phase at the receiving end within the same feedback circuit.

10.3.3.4 Differential decoder and symbol to byte mapping

After QAM-demodulation and pulse reshaping, the phase state, which corresponds to a certain quadrant (see Table 10.3), is provided to a

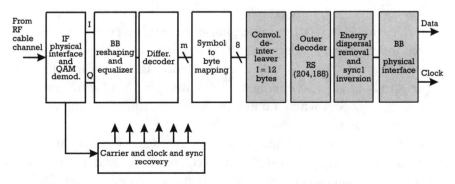

Figure 10.13 Conceptual decoding system description.

differential decoder. The output of the decoder delivers the corresponding two MSBs of the m bits symbol. Next, the m bits symbols are processed in order to regain the original symbols with a length of 8 bits each. The required synchronization for this process is enabled by the synchronization pulse in the TS.

10.4 DVB terrestrial

Television signals can be provided via terrestrial networks as well. This section first discusses the basic elements of terrestrial communications, then describes the several DVB systems for terrestrial communications at a functional level, and finally explains the specifications for the channel encoding and decoding process.

10.4.1 Elements of terrestrial communications

In case of terrestrial communications, specific elements play an important role. These elements concern typical transmission characteristics which have to be regarded when designing a terrestrial communication system.

10.4.1.1 Transmission medium

Digital terrestrial television services are expected to be provided via the *ultra high frequency* (UHF) band. The frequencies in this band range from 0.3 GHz to 3 GHz.

In contrast with satellite and cable systems, the terrestrial transmission of signals often suffers from multipath interference. A broadcast signal can be reflected, for example, by high buildings or mountains. The reflections are added to the main signal at the receiving end. Because the reflections travel via a different (and thus longer) route, these signals are delayed and therefore are called *echoes*. Hence, multipath interference occurs. Additionally, depending on the power used, cochannel interference may be caused when a different station transmits its programs via the same frequency.

In cities with high buildings and in mountain areas, echoes are likely to appear. When terrestrial signals are received by a fixed antenna, the antenna can be aimed at the strongest (main) signal. Hence, the influence of echo signals is minimized. This channel can be seen as a Rice channel. The Rice channel is described by the main signal and the sum of all echo signals together. However, in the case of portable reception, the power of the main signal drops more or less to the same power level as that of the echo signals. The channel can now be considered a Rayleigh channel, which is described by the sum of all (delayed) signals received.

10.4.1.2 Spectrum efficiency

For maximum spectrum efficiency within the UHF band a *single frequency network* (SFN) operation can be used. A SFN is built up out of broadcast stations which simultaneously transmit identical data streams via the same frequency. Neighboring broadcast stations support each other in their function. Moreover, if a large distance between neighboring broadcast stations is possible, national coverage can be achieved.

Beside the power level used, the distance between the broadcast stations mainly depends on the length of the guard interval. A relatively long interval allows a larger distance. For example, a guard interval of 200 μs corresponds to a distance of 60 km (200 μs*300,000 km/s = 60 km). The spectrum efficiency can be tailored to specific requirements by a flexible guard interval.

Terrestrial systems are designed for transmission via the same medium as satellite systems. Correspondingly, frequencies can be reused by the application of orthogonal polarization.

10.4.1.3 Modulation

In correspondence with digital satellite and cable systems, M-ary signaling is used for digital terrestrial services. Depending on the specific requirements, QPSK or QAM can be used. In case of QPSK energy dispersal scrambling has to be applied to avoid adjacent channel interference.

10.4.2 DVB terrestrial systems

DVB has specified several systems for terrestrial communication. Besides the digital terrestrial system, DVB specified a *multipoint video distribution system* (MVDS) and a system for *microwave multipoint distribution service* (MMDS). These systems are discussed at a functional level with their typical application and characteristics.

10.4.2.1 Digital terrestrial system

In general, terrestrial systems can provide local and national coverage in a more cost-effective way than satellite and cable systems. Moreover, the introduction of digital terrestrial systems enables a dramatic increase in available frequency spectrum. These frequencies, for example, can be used for the growing demand for mobile communications.

The DVB draft specifications (prETS 300 744 [10]) for a digital terrestrial system allow stationary and static portable reception via 8-MHz channels in the UHF band. Furthermore, these specifications include the use of large-area SFNs to allow maximum spectrum efficiency.

Different starting conditions in individual countries may lead to different introduction scenarios of digital terrestrial systems. This, for example, can depend on the spectrum availability and the number of existing analog services.

10.4.2.2 Multipoint video distribution system

The DVB specifications (ETS 300 748 [11]) for the MVDS are compatible with the 11/12 GHz satellite system (ETS 300 421) using QPSK modulation. The actual difference lies in the frequency band used for the transmission of digital terrestrial signals. Although the MVDS typically operates in the frequency band 40.5 to 42.5 GHz, the system is applicable to other frequency bands above 10 GHz. Moreover, the MVDS is suitable for use on different transmitter bandwidths varying from 26 MHz channels to 54 MHz channels. The frequency spectrum of the adjacent channels overlap each other in part but can be separated by the use of orthogonal polarization.

Typically, the MVDS is applied in areas where no cable system is provided. Moreover, it can be a competitive alternative for cable systems. An MVDS consists of (omni)directional transmitters and a number of stationary receivers. The maximum broadcasting distance of a digital MVDS is 6 km [12]. If one operator has been allocated the full 2-GHz frequency spectrum and the use of four broadcasting stations, 120 to 384 digital television programs can be received in an area of 200 km^2.

10.4.2.3 Microwave multipoint distribution service

At this moment DVB is working on the specifications of an MMDS (prETS 300 749 [13]). This digital terrestrial system is compatible with the DVB cable system (ETS 400 429). Hence, it uses 8-MHz terrestrial channels, and QAM modulation (16-QAM, 32-QAM, and 64-QAM) is applied. By using 32-QAM, a bit rate compatible with terrestrial *plesiochronous digital hierarchy* (PDH) can be retransmitted in an 8-MHz channel as well. The MMDS operates at frequencies below 10 GHz.

Analog to MVDS, this system is also typically applied as an extension of the CATV network and can serve as an alternative for cable systems in rural areas. Moreover, MMDS is perfectly suited to provide digital terrestrial television services within buildings with a large number of subscribers. This shows much correspondence with the digital SMATV system.

10.4.3 Channel encoding

This section describes the DVB specifications for the digital terrestrial system, explaining how the constraints in the basic elements of terrestrial communication (see Section 10.4.1) are respected by DVB in the specifications for channel encoding.

10.4.3.1 Encoding system

The specifications for MVDS and MMDS are the same as the DVB standards for digital satellite and cable systems. The specifications for the DVB terrestrial system, which are compatible with the DVB digital satellite system, are explained here. Figure 10.14 shows a conceptual description of the digital terrestrial system. The elements of this system which are common to the digital satellite encoding system are represented in gray.

The most important steps to adapt the TS to transmission via a terrestrial channel are the following.

▶ Transport multiplex adaptation;

▶ Randomization for energy dispersal;

▶ Outer error correction coding and outer interleaving;

▶ Inner error correction coding and inner interleaving;

▶ Mapping and modulation;

▶ OFDM transmission.

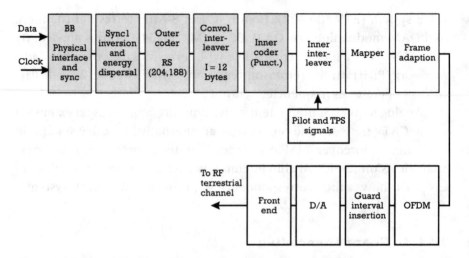

Figure 10.14   Conceptual encoding system description.

10.4.3.2   Inner interleaving, symbol mapping, and modulation

After the inner error correction procedure, the TS is demultiplexed into several substreams. When QPSK ($m = 2$) is used to eventually modulate the carriers, the data stream is demultiplexed into two substreams (see Figure 10.15). When 16-QAM ($m = 4$) or 64-QAM ($m = 6$) is used, it results in four or six substreams, respectively. Next, inner interleaving (bit-wise interleaving and symbol interleaving) is applied. A symbol is formed by the outputs of the m bit-wise interleavers. Hence, each symbol consists of exactly one bit from each of the m bit-wise interleavers. The purpose of the symbol interleaver is to map the m bit symbols onto the carriers. Finally, the carriers are QPSK, 16-QAM, or 64-QAM modulated and transmitted.

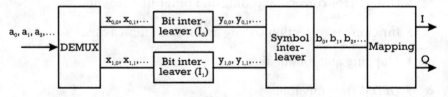

Figure 10.15   Mapping of input bits into QPSK modulation symbols.

10.4.3.3 OFDM transmission

In case of multipath interference, the delay of the echo is often longer than the symbol duration of the main signal. This results in a high level of interference. Echoes can be countermeasured by making the symbol duration longer. In turn, this would lead to more required bandwidth. However, a tradeoff between bandwidth and symbol duration is possible.

A method of achieving a larger symbol duration within the same bandwidth is to demultiplex a distinct symbol into several subsymbols. Next, the subsymbols are modulated in parallel onto different carriers. The total bandwidth (sum of all carrier frequencies) remains the same. Hence, the subsymbol duration is increased. Next, the modulated subsymbols are added, after which the newly obtained data stream can be transmitted.

DVB has chosen *orthogonal frequency division multiplex* (OFDM) as the technology for transmitting digital signals via the digital terrestrial system. OFDM is a multicarrier transmission technology that is currently being used in *digital audio broadcasting* (DAB). Typically, all adjacent carrier frequencies are orthogonally polarized.

The OFDM transmission system specified by DVB is able to operate in a 2k mode and 8k mode. In case of the 2k mode a maximum of 1,705 carriers per OFDM symbol can be used. The 8k mode is specified for a maximum of 6,817 carriers per OFDM symbol. The symbol duration in the latter is longer. Hence, a larger transmitter distance is allowed. Both modes are suitable for single transmitter operation. Furthermore, the 2k mode can be used in small SFN with limited transmitter distance. The 8k mode can be used in either large or small SFN.

10.4.3.4 OFDM frame structure

The OFDM signals are organized in a frame structure. Each frame consists of 68 OFDM symbols. Four frames together constitute a super-frame. As each symbol in its turn is modulated on a number of carriers, a matrix arises. The distinct elements of the matrix are referred to as cells. DVB has specified 1,512 active carriers for the 2k mode. In case of the 8k mode the number of active carriers is 6,048. The rest of the carriers is formed by reference data (i.e., scattered pilot cells, continual pilot cells, and *transmission parameter signaling* (TPS) carriers (see Table 10.4)).

By the application of pilot cells, frame synchronization, frequency synchronization, time synchronization, channel estimation as well as

Table 10.4
OFDM Frame Structure

Parameter	2k Mode	8k Mode
Maximum carriers	1,705	6,817
Active carriers	1,512	6,048
Scattered pilot cells	131	524
Continual pilot cells	45	177
TPS carriers	17	68

transmission mode identification are established. The pilot cells are always transmitted at a higher or "boosted" power level. (For example, in the case of QPSK, the amplitude is raised by a factor of 2.) The TPS carriers contain information concerning the applied channel coding and type of modulation.

10.4.3.5 Guard interval insertion

The DVB specifications include the use of a flexible guard interval between adjacent channels. A relatively long guard interval increases the transmitter distance but reduces the bit rate capacity (i.e., the symbol duration is longer). A flexible guard interval thus allows a tradeoff between transmitter distance and bit rate capacity. For the 8k mode with $T_g = 224 \, \mu s$, this results in a maximum transmitter distance $d_{t,\max} = 67$ km (relevant for national coverage). In case the code rate is 7/8, this corresponds to a bit rate of 26.1 Mbps [14]. In Table 10.5, the guard interval related to the maximum transmitter distance and bit rate is presented for the 8k mode and 64-QAM is applied.

Table 10.5
Relation Between Guard Interval, Maximum Transmitter Distance, and Bit Rate

8k Mode		
T_g [μs]	dt,max [km]	bit rate [Mbps]
224	67	26.13
112	33.5	29.03
56	16.8	30.74
28	8.4	31.67

Notes: Code rate = 7/8; 64-QAM.

10.4.3.6 Hierarchical coding

CATV networks generally make use of coax cables. This results in minimal external interference at higher frequencies. When the CATV network topology is configured adequately, the transmission quality of the network can be considered constant and high. In case of satellite and terrestrial transmission external interference can be caused by rain if both polarizations are used. A digital satellite system (incorporating error correction) either performs at the required level or, when the external interference exceeds a certain threshold, the digital signal is interrupted. This can (partly) be countermeasured by accurately directing the antenna towards the satellite, a satellite dish with a larger diameter, or both.

The terrestrial transmission quality depends on local characteristics. A transmitter may, for example, cover a whole city, but because of obstruction the transmission quality in a lower located city area can be below the required level. In case of digital transmission this could lead to the interruption of the signal. Countermeasures are, for example, raising the transmitter power level considerably, the allocation of an extra transmitter, or the application of QPSK rather than 16-QAM or 64-QAM. Another possibility is the use of a lower code rate (i.e., more error correction information is added at the cost of a higher bit rate). These countermeasures, however, lead to an increase in costs or a decrease of the total transmission quality to serve only a fraction of the home users.

The solution for the problem described above is the application of hierarchical coding. DVB has specified two-level hierarchical channel coding. Technically, this implicates a "splitter" separating the incoming transport stream into a high-priority and low-priority transport stream. Both streams undergo their own inner/outer error correction coding process and inner/outer interleaving process. Next, these two bit streams are provided to the input of the mapper, after which modulation takes place (see Figure 10.16). Hierarchical coding is applied in case of 16-QAM and 64-QAM only.

DVB has specified the high-priority stream with a high code rate, and thus results in a low bit rate. For the low-priority stream a low code rate is specified, which results in a high bit rate. Hence, a low bit rate, rugged version or a high bit rate and less rugged version of the same program can be received. It is also possible to transmit entirely different programs on both separate streams. This requires the provision of hierarchical source coding. The high-priority stream could, for example, be used for a normal

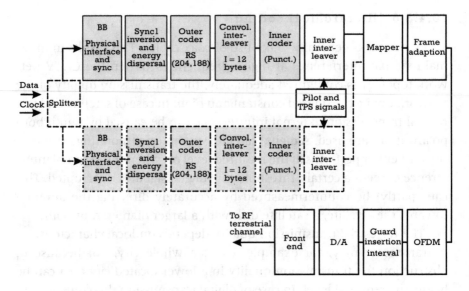

Figure 10.16   Hierarchical channel coding.

program, while the low-priority stream could be applied for the same program with HDTV-quality. DVB has decided not to adopt hierarchical source coding.

Hierarchical channel coding leads to a different constellation diagram. A high-priority stream makes use of a constellation with higher amplitudes. Three different levels (alpha = 1, 2, or 4) are specified by DVB. Figure 10.17 describes a 16-QAM constellation diagram with alpha = 2. An increase of alpha (i.e., a higher amplitude) implies a higher output power level of the transmitter. However, in case the transmitter power is constant and the value of alpha increased, the channel isolation has to be increased in order to maintain the same modulation quality. Moreover, a higher value of alpha results in more influence of phase noise. For each OFDM symbol the value of alpha and code rate are included in the TPS carrier.

10.4.4   Channel decoding

Section 10.4.3 explains the DVB encoding system for the digital terrestrial system. At the receiving end, the signal has to be decoded. This section discusses the DVB specifications for the required channel decoding.

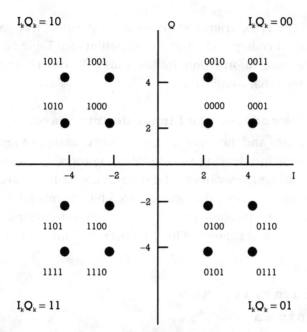

Figure 10.17 Constellation diagram 16-QAM and alpha = 2. (*$I_k Q_k$* are the two MSBs in each quadrant.)

10.4.4.1 Decoding system

In spite of the fact that the DVB specifications for the digital terrestrial system are in the final stage of becoming an ETSI standard, this system still has to evolve from the laboratory environment to a practical implementation. Tests have shown that portable reception in a car driving at 170 km/h is possible. This could make the system very interesting for all sorts of mobile digital broadband video services.

However, especially in the case of the 8k variant, the required sophisticated decoder technology is not yet being produced in mass production. Depending on the actual costs of the decoder's practical implementation, the requirements for the digital terrestrial system may still be subject to change.

10.4.4.2 Recovery of Reference Information

With the application of pilot cells, synchronization (frame, frequency, and time), channel estimation, and transmission mode identification are established. The TPS carriers contain specific information concerning the applied channel coding and type of modulation (see Table 10.6).

The reference information (pilots and TPS) can be recovered by means of a feedback circuit (PLL).

10.4.4.3   Demodulator and inner de-interleaver

The demodulator and the inner de-interleaver operate in a reversed way compared to the interleaving and modulation process at the transmitting end. The (high- and low-priority) bit streams are de-interleaved in order to be demodulated. Next, the demodulated bit streams are multiplexed into a single bit stream again in order to be applied to the inner decoder. The whole process is supported by the reference information.

10.5   Summary and conclusions

DVB has provided specifications for a broad variety of digital transmission systems concerning communication via satellite, CATV, and terrestrial networks. These systems have been or are currently being standardized by ETSI. Table 10.7 provides an overview of the typical parameters of the DVB digital transmission systems.

Satellite communication suffers from power limitations. For this reason, QPSK is used as a modulation method. The advantage of satellite communication is the rich availability of bandwidth, which in the end allows a high bit rate. After receiving the satellite signals, further transmission via a SMATV network is enabled by the DVB SMATV system. This allows the use of QAM, but the available bandwidth is limited. Hence, a

Table 10.6
TPS Carrier Information

Constellation including the value of alpha (QAM modulation)
Hierarchy information including inner code rate
Guard interval
Transmission mode (2k or 8k)
Frame number in a super frame
Synchronization word

Table 10.7
Typical DVB Digital Transmission System Parameters

DVB System	Modulation	Frequency Band (GHz)	Signal Bandwidth (MHz)	Bit Rate (Ru) (Mbps)
DTH	QPSK	11/12 (downlink)	26.0–54.0	18.7–68.0
SMATV	16-QAM	0.23–0.47 or 0.95–2.05	5.9–7.9	18.9–25.2
	32-QAM	0.23–0.47 or 0.95–2.05	4.7–8.0	18.9–31.9
	64-QAM	0.23–0.47 or 0.95–2.05	3.9–8.0	18.9–38.1
CABLE[1]	16-QAM	f	2.0–7.9	7.0–27.3
	32-QAM	f	2.0–8.0	8.7–34.6
	64-QAM	f	2.0–7.9	10.4–41.3
TERRESTRIAL[2]	OFDM (QPSK)	0.3–3.0	7.6	5.0–10.6
	OFDM (16-QAM)	0.3–3.0	7.6	10.0–21.1
	OFDM (64-QAM)	0.3–3.0	7.6	14.4–31.7
MVDS	QPSK	40.5-42.5	26.0–54.0	18.7– 68.0
MMDS[3]	16-QAM	f<10	2.0–7.9	7.0–27.3
	32-QAM	f<10	2.0–8.0	8.7–34.6
	64-QAM	f<10	2.0–7.9	10.4–41.3

Notes:
1. Frequency band is chosen by CATV operator;
2. Nonhierarchical coding;
3. Frequency band has not yet been allocated to MMDS.

lower bit rate is achieved. This last case more or less applies to the DVB cable system as well.

As a result of the different local characteristics, the DVB terrestrial system is the most complicated system of all. Multipath interference is countermeasured by means of the OFDM transmission technology. This multicarrier solution allows the application of QPSK as well as QAM. Due to the limited terrestrial frequency spectrum, this system has a lower maximum bit rate than the DVB DTH system.

MVDS, a terrestrial system that, for example, can be chosen as an alternative for transmission via CATV networks, is identical to the DVB DTH system. The only difference is the frequency band in which this system is operated. This enables a compatible use of both systems. The MMDS system provides the user with the same functionality as the MVDS. The system design, however, is based on the DVB cable specifications. This makes the DVB MMDS system compatible with the DVB cable system.

References

[1]   Reimers, U., *Digitale Fernsehtechnik, Datenkompression und Übertragung für DVB,* Springer, April, 1995.

[2]   ITU, *Radio Regulations,* 1990 edition, revised in 1994, Geneva, 1994.

[3]   Ha, T. T., *Digital Satellite Communications,* New York: Macmillan Publishing Company, 1988, pp. 25–28.

[4]   EBU/ETSI JTC, *Digital broadcasting systems for television sound and data services; Framing structure, channel coding and modulation for 11/12 GHz satellite services,* ETS 300 421, December, 1994.

[5]   EBU/ETSI JTC, *Digital broadcasting systems for television sound and data services; Satellite Master Antenna Television (SMATV) distribution systems,* ETS 300 473, May, 1995, p. 8.

[6]   EBU/ETSI JTC, *Digital broadcasting systems for television sound and data services; Satellite Master Antenna Television (SMATV) distribution systems,* ETS 300 473, May, 1995.

[7]   ISO/IEC DIS 13818-1, *Coding of moving pictures and associated audio,* June, 1994.

[8]   Viterbi, A. J., *Error Bounds for Convolutional Codes and an Asymptotically Optimum Decoding Algorithm,* IEEE Trans. On Information Theory IT-13, No. 2, 1967.

[9]   EBU/ETSI JTC, *Digital broadcasting systems for television sound and data services; Framing structure, channel coding and modulation for cable systems,* ETS 300 429, December, 1994.

[10]   EBU/ETSI JTC, *Digital broadcasting systems for television sound and data services; Framing structure, channel coding and modulation for digital Terrestrial television (DVB-T),* prETS 300 744, November, 1996.

[11]   EBU/ETSI JTC, *Digital broadcasting systems for television sound and data services; Framing structure, channel coding and modulation for MVDS at 10 GHz and above,* ETS 300 748, October, 1996.

[12]   TNO FEL, *Inventarisatie van MVDS systemen ten behoeve van beleidsvorming door HDTP,* maart 1996, p. 35.

[13] EBU/ETSI JTC, *Digital broadcasting systems for television sound and data services; Framing structure, channel coding and modulation for MMDS systems below 10 GHz,* prETS 300 749, 11 January, 1996.

[14] EBU/ETSI JTC, *Digital broadcasting systems for television sound and data services; Framing structure, channel coding and modulation for MMDS systems below 10 GHz,* prETS 300 749, 11 January, 1996, p. 40.

Contents

Conditional access

11.1 Introduction

In general, a CA system is a system that provides access to users when specific requirements are met. These requirements can, for example, refer to identification, authentication, authorization, registration, payment, or a combination. One of the technical means to prevent unauthorized users to get access to services is encryption. In the context of pay-TV, a CA system ensures that only authorized users (i.e., users with a valid contract) can watch a particular programming package [1]. In technical terms, a television program is broadcast in encrypted form and can only be decrypted by means of a set-top box. The set-top box incorporates the necessary hardware, software, and interfaces to select, receive, and decrypt the programs.

Chapter 3 discusses the interests of the several parties involved in CA. This chapter details the technical aspects of CA, explaining the basic elements of a CA system—encryption, key management, subscriber authorization, and subscriber management—and discussing the DVB specifications for the common scrambling algorithm (CSA) at a functional level. The CSA was developed to encrypt programs in a uniform way. This uniform encryption algorithm forms the basis of three different models of CA. In the *Multicrypt* model, a common interface allows a multitude of different service providers to make use of the same set-top box, in which the CSA is implemented. Within the *Simulcrypt* model, the CSA allows the same transponder channel to be shared by different service providers. This allows the same program to be received by both service providers' set-top box populations simultaneously. Finally, in case of the *Transcontrol* model the CSA allows CATV operators to control (i.e., to manage) the services that are provided via their network by other CA providers. The models for Multicrypt, Simulcrypt, and Transcontrol as well as the applicable DVB specifications are discussed in Sections 11.4, 11.5, and 11.6 respectively.

11.2 Elements of conditional access

Several elements play a role in CA systems used for pay-TV services. This section first discusses the important building blocks concerning encryption and key management and then focuses on more specific elements such as subscriber management and subscriber authorization, which together construct the actual CA management system.

11.2.1 Encryption

The main function of a CA system is to ensure that only authorized users can watch and/or hear a particular programming package. For this purpose the audio and video signals are processed in such a way that the program concerned cannot be viewed and heard in a normal way. This process is referred to as scrambling. For the scrambling of analog signals, analog techniques (e.g., the addition of interfering carrier waves, the modification of synchronization or color burst information, and the delay or inversion of the video signal) as well as digital techniques (e.g., cut and rotate and line shuffling) can be applied.

Cryptography is commonly used as a technique to encipher (i.e., scramble) and decipher (i.e., descramble) information. Enciphering and deciphering are processed by means of algorithms and keys. An algorithm can be regarded as the program by which enciphering and deciphering are realized. A key (also called control word) provides access to this program. The combination of the algorithm and the key forms the crypto system [2]. In case cryptography is applied in the digital domain, the terms encryption and decryption are used as equivalents of the terms enciphering and deciphering, respectively.

With many crypto systems the same key is used for enciphering as for deciphering. This key is known to the sender and to the receiver of the information only, and has to remain secret. Such systems are called symmetric crypto systems, because communication in both directions is possible due to shared knowledge and application of one and the same secret key. The secret key will preferably have to be exchanged via an alternative to the communication channel. The secret key can, for example, be exchanged via mail.

The 1976 article "New Directions in Cryptography" by Diffie and Hellman [3] initiated a new development, namely that of a public crypto system. They introduced a model that included a unique pair of keys for the sender as well as for the receiver. A key pair consists of a secret and a public key. Both parties know the content of the other party's public key. Over a year later Rivest, Shamir, and Adleman [4] published their model for a public crypto system. The name of the system developed by them is abbreviated to RSA, representing the first letters of the makers' last names. The typical quality of this model is that when information is enciphered with the secret key of the key pair, the enciphered information can be deciphered with the public key of the same key pair and vice versa. Furthermore, it is not feasible to deduct the secret key from the public key and the other way around.

So when Alice sends a message to Bob and the message is enciphered with Bob's public key, then only Bob is capable of deciphering this message with his secret key. This way, the secrecy or confidentiality of the message is guaranteed. Also, it is possible for Alice to encipher the message with her secret key. Bob can decipher this enciphered message by using Alice's public key. This way Bob can be certain that Alice is the one who sent the message. Alice has put her digital signature on the message, so to speak. Because different keys are required for the enciphering and deciphering, this is an asymmetrical crypto system. The great advantage of this system is that it is not necessary to exchange a secret key preceding the communication.

At first sight it may seem illogical, but in cryptography it is assumed that everyone knows the cryptographic algorithm (Kerckhoff's principle). By keeping a new algorithm secret, the security is only ensured for a short period of time. This can be attributed not only to leaks in the organization, but sometimes also to the ingenuity of the computer criminal. Computer criminals are often able to find totally different ways to break the encipherment. Thus, in cryptography the keys are the central element. If secret keys become public, the crypto system's security can no longer be guaranteed.

11.2.2 Key management

If encryption is used in digital communication, the communicating parties must make a number of procedural agreements in advance. Beside an algorithm and a method of application and initialization, there has to be an agreement on keys. Key management contains every aspect of the handling of keys. Thus, it contains generation, distribution, storage, replacement/exchange, use, and the eventual destruction of keys [5].

A symmetrical system is complicated by the difficulty of a secret key that has to be chosen and then made available to the parties at both ends of the communication line. In most cases this exchange is affected by encipherment of the secret key with another key, etc. Eventually a key will have to be distributed along a different kind of communication line—for example, using a special courier. Nevertheless, the confidentiality of the particular key(s) is of eminent importance.

In case of asymmetrical systems, only the public keys have to be exchanged by the communicating parties. The crypto system cannot be compromised by the disclosure of the public keys. However, this introduces the problem that it is impossible to determine that the public key is actually the public key of the party with which communication is intended. An independent third party, or *trusted third party* (TTP), can acknowledge the connection between the public key and its user by issuing a certificate [6]. This certificate primarily binds an identity to a public key and is enforced with the digital signature of the TTP. Hence, the TTP acts as a *certification authority*. The user gives a copy of the certificate to the other communicating party. This can also take place directly through the TTP. A different possibility is that a TTP provides the public with certificates by using a directory service. This directory can be regarded as a phone book.

Additionally, a TTP can provide key management services. These services can, in contrast with the current practice, also be applied in CA systems. A TTP could, for example, generate and distribute cryptographic keys, without having control over the CA system itself. In this respect a TTP, rather than the CA service provider, takes care of the key management part of the CA system.

11.2.3 Conditional access management systems

In addition to the facility to encrypt and decrypt the programs concerned, the CA system consists of a CAMS. A CAMS provides network management and in turn comprises a *subscriber management system* (SMS) and a *subscriber authorization system* (SAS). The SMS is an administrative system that stores customer data and requests and that eventually issues invoices. The SAS is the technical managing system that processes the data from the SMS into commands that can be interpreted by the set-top box.

In the current situation vertically integrated service providers operate on the market. These service providers not only package programs and sell or rent set-top boxes, but they also control the CAMS. They are dependent on their canvassed subscribers, who pay their monthly subscription. Constructing a clientele requires a great effort. It is therefore important to these service providers not to make the market (i.e., the client data) easily accessible to other providers. Moreover, the confidential handling of this data is also in the interest of individual users as well. Hence, the control over the SMS is crucial for the information service provider. From this perspective, the control over the SAS, on the other hand, is less crucial.

In contrast with the current practice, the SAS can also be controlled by another party, for example, network service providers such as CATV operators [7]. This implies that the commands of the SMS have to be passed to the CATV operators' SAS. To avoid the revelation of client data to the CATV operator, this data can be made anonymous, while one is still able to process the network management. However, vertically integrated service providers are not very keen on not having the authorization control over the CA system. They want to be able to directly start or stop the service provision depending on the timely payment of the subscription fee.

Technically, every set-top box derives the same data from a common control word for the decryption process. This implies that the keys and algorithms, which are required to access a particular service, have to be present in every set-top box. These keys are called the operating keys or session keys. The architects of CA systems try to protect the security of the CA system by frequently changing the session keys. Hence, the life cycle of these keys is shortened and the risk that a key will be found is reduced. The operating keys have to be distributed in a secure (i.e., enciphered) way. A management key is used to encrypt the session keys. The management key can be present in one or in a group of set-top boxes. In the latter case, the management key is called the group key. Depending on the complexity of the CA system, several management keys that, in turn, have to be distributed securely by means of other keys may exist. In the end, at the highest level, only one key, called the unique key, exists. This key is present in one set-top box only and is related to the unique address of this set-top box.

11.3   DVB common scrambling algorithm

DVB specified a crypto system for use in CA systems. This section discusses the operation of this system at a functional level. Moreover, the several distribution agreements, which apply to the specific elements of the DVB crypto system, are explained.

11.3.1   DVB crypto system

Canal+ SA, Centre Commun d'Etudes de Telediffusion et Telecommunications, Irdeto BV, and News Datacom Limited, after consultation with several export control-related authorities, have specified the DVB crypto system for CA to be applied within digital broadcasting systems [8]. This system is referred to as the common scrambling algorithm. In fact, the common scrambling algorithm is comprised of the *common descrambling system* and *scrambling technology*, rather than the algorithm only. Moreover, DVB uses the term scrambling instead of encryption. To avoid any confusion about the terminology and the difference between the CSA and an actual algorithm, this section uses the terms encryption and decryption. The term algorithm is used for the program by which encryption and decryption are processed. Figure 11.1 presents a functional description of the DVB encryption system.

By means of a demultiplexer the data stream of the TS (or at PES level if required) is separated into two kinds of data, the data that must be encrypted and the data that must not be encrypted. Because the header of the TS contains, among other things, information for synchronization, it must not be encrypted. Otherwise it is no longer feasible to synchronize at the receiving end. Moreover, a service may be provided for free (free-TV), in which case encryption of the concerned data is not needed. In case a PES is used, certain constraints are applicable:

▶ Encryption only takes place at either TS level or at PES (*packetized elementary stream*) level. Moreover, it is not allowed to encrypt at both levels simultaneously.

▶ The header of a PES packet must not exceed 184 bytes.

▶ The TS packets, which carry parts of an encrypted PES packet, do not have adaptation fields. The only exception are the TS packets containing the end of a PES packet. To align the end of the PES packet with the end of the TS packet, the TS packet carrying the end of an encrypted PES packet may carry an adaptation field.

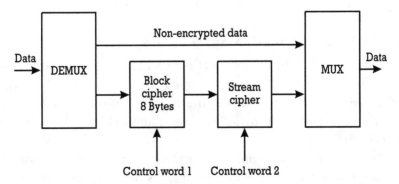

Figure 11.1   Functional description of DVB encryption system.

The encryption of data is processed by a block cipher and a stream cipher, respectively. The block cipher encrypts the data stream per 8 bytes. The stream cipher encrypts the data stream bitwise. The algorithm is designed in such a way that the memory needed for the decryption process is considerably less than that needed for the encryption process. This allows a low-cost set-top box at the expense of a complex and thus more expensive encryption system. The exact amount of memory depends on the actual implementation of the crypto system.

Both ciphers are executed by means of different control words. These control words are encrypted with a proprietary encryption system and included in the SI, which, in turn, is included in the header of the TS. The encrypted control word used for the block cipher is called the *entitlement control message* (ECM) and the one used for the stream cipher is referred to as the EMM. An EMM authorizes the smart card concerned to receive the television program and an ECM is used to let an authorized smart card decrypt the program. The television program is thus decrypted if the smart card is authorized by an EMM and if it receives the corresponding ECM. In technical terms this means that the original control words are derived from the EMM and ECM, after which the decryption can be executed.

The EMMs are related to the customer data contained in the SMS. The SMS controls whether and for which period a customer is entitled to receive television programs. For pay-TV the EMM may be changed each month or day. In case of pay-per-view or video-on-demand the EMM is changed per program.

According to the European Union Directive on Television Standards [9], the use of the CSA in consumer equipment for CA is mandatory.

11.3.2 DVB distribution agreements

The four companies that developed the CSA agreed to license this system. As a contribution to DVB, they also agreed that a low nominal royalty is to be charged to each licensee to keep the price of a set-top box as low as possible. However, different royalties may be applicable to the several elements of the CSA.

The *DVB Common Scrambling Algorithm Distribution Agreements* [10] have been developed by the four companies. These agreements include, among others, the *Common Descrambling System License Agreement* for the manufacturers of set-top boxes and their components, providers, design-ers and other entities engaged in CA. For the manufacturers of scram-blers the *Scrambling Technology License Agreement* applies. In turn, these manufacturers sublicense the purchasers of scramblers. In contrast with Kerckhoff's principle, the DVB crypto system (including the actual algo-rithm as well as the key length) is kept confidential. For this purpose com-panies (that have not engaged in audiovisual piracy) have to sign the *Non-Disclosure Agreement* as well.

The Directive on Television Standards requires that the CSA is admin-istered by a recognized standardization body. The *DVB Descrambling Custo-dian Agreement* and the *DVB Scrambling Technology Custodian Agreement* specify these terms of administration with ETSI acting as neutral custo-dian for the CSA specifications.

As a result of the ETSI organization being based in France, companies were concerned that, in addition to the licensing procedures described above, French companies would have a competitive advantage by not having to apply for a French export control license. Because France also requires a license for the use of cryptographic goods, in the end, the level playing field remains intact.

11.4 Multicrypt

Beside the CSA, a CA system includes a CAMS(conditional access man-agement system). This section explains, among other things, why the CAMS is not standardized. Hence, a completely standardized set-top box is not achieved. To avoid a situation in which a consumer ends up with many set-top boxes from different service providers, DVB decided to sup-port the *Multicrypt* model. The model and its technical implementation are discussed in Section 11.4.1 and 11.4.2 respectively.

11.4.1 The Multicrypt model

In most situations, a vertically integrated entity operates on the market. This entity not only packages programs and sells or rents set-top boxes to its subscribers, but also controls the complete CAMS. In case other service providers, using different transponder channels and different CAMSs, want to make use of the same set-top box, these CAMSs have to be interpretable by the set-top box as well. Because of efficiency and cost effectiveness, this requires a common encryption algorithm to be present in the set-top box, rather than several algorithms. Additionally, a common interface between this common encryption part and the various CAMs is needed to facilitate a multitude of service providers. This model therefore is referred to as Multicrypt.

11.4.2   DVB common interface

DVB decided not to standardize the CAMS. An important reason is that when the CA system has been subject to piracy, the security of all similar CA systems (also in other countries) can be compromised. Moreover, this implies that rights to broadcast in certain geographical areas can no longer be protected. Hence, programs could be broadcast without prior authorization. Another important reason for not standardizing the CAMS is that the investments of current service providers in their proprietary CAMSs, particularly the SMS that contains the client data, would be undermined. As a result of not standardizing the CAMS, a common interface is needed to make the proprietary CAMSs and the program, which is encrypted with the standardized CSA, interoperable. As such, a common interface introduces flexibility rather than extra functionality in respect to watching programs.

DVB has specified a standard [11] for a *common interface* (CI). The CI forms a standardized interface between a host and a module. A host is a device in which one or more modules can be connected (e.g., an IRD, a television set, a PC, or a VCR). In this context a module is a device, operating only in combination with the host, designed to process specialized tasks in association with the host (e.g., a CA subsystem or an electronic program guide (EPG). This allows service providers to choose solutions from different suppliers for their systems within the specifications, and this, in turn, provides freedom of choice for antipiracy technologies. Moreover, by the application of the CI, the consumer only needs one host for watching programs from different service providers. Figure 11.2 shows an example of a CI, using an external module. If required, proprietary functions can be implemented in this external module.

A tuner within the host is used to tune in on the required channel. This signal is then demodulated, providing a scrambled MPEG-2 TS

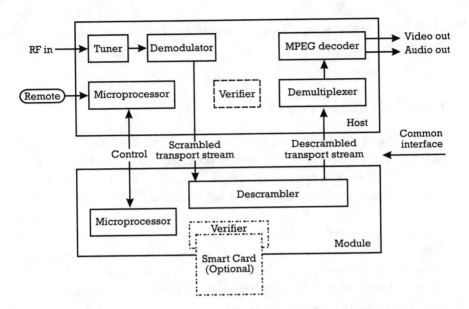

Figure 11.2 Common interface between host and module. (*Source:* DVB project.)

MPEG decoder. Ultimately, this process supplies a separated audio and video signal.

The CI itself consists of a logical interface and a command interface. The former concerns the MPEG-2 TS. The latter provides the control information between the host and module. The existing *Personal Computer Memory Card International Association* (PCMCIA) II standard has been used for the physical connection between the host and the module. This standard, which is applied in the personal computer industry worldwide, specifies the physical connection and the logical functions the card must execute. However, the implementation of the PCMCIA card is not mandatory. Thus, connecting an extra smart card system to the module is possible. In this case a verification system is required in the host as well as in the module.

By means of the Directive on Television Standards the CI is mandatory for television sets with a built-in digital signal decoder (i.e., built-in host). The CI is not made mandatory for use in set-top boxes. This implies that a digital set-top box, which is capable of decrypting encrypted DTV signals, does not have to meet the specifications of the CI, while the use of the CSA in the set-top box is mandatory! This option is thus left to the market.

11.5 Simulcrypt

Until now, vertically integrated service providers have protected their (geographical) markets by not giving other service providers access to their proprietary CA system. However, in case other service providers wish to provide the same service in a different geographical area, there is no direct competition. When the vertically integrated entity provides its services via a satellite, the expensive transponder channel could be shared in such a way that this service is provided to the set-top box populations in the different geographical areas of both service providers. This requires the use of *Simulcrypt*. In this section, the Simulcrypt model and a typical implementation are discussed.

11.5.1 The simulcrypt model

It may occur that a service provider broadcasts the same program at the same time to its subscribers in a different geographical area than the vertically integrated provider. This may, for example, be relevant with a total programming package or with a live popular sporting event. In this situation it would be efficient and cost effective to share the same transponder channel to broadcast the same program to the two providers' subscribers. This not only requires a common encryption algorithm, but the control information of the requesting providers' CAMS needs to be accommodated in the broadcast signal as well. Hence, a single smart card or CA module and a single set-top box are necessary to access the local service. Because the same program can be received by both service providers' set-top box populations simultaneously, this model is also known as Simulcrypt.

The application of Simulcrypt requires commercial negotiations between the service providers concerned. Beside the situation described above, it may be that a service provider wishes to use the DTV services of the vertically integrated entity on a continuous basis, while using another transponder channel. Hence, a competing program may be provided via the users' set-top box in the same geographical area at the same time. Because of the provision of a competing service and the absence of cost effectiveness as a result of the sharing of a transponder channel, in the latter case the negotiations are likely to be more difficult and may be theoretical.

After intensive debate DVB decided to develop a *Code of Conduct* that allows service providers to make use of the DTV services of a CA service provider on a nondiscriminatory basis. This Code of Conduct is adopted in the Directive on Television Standards. About one and a half years after the establishment of this directive, DVB produced the technical specifications (TS101 197-1 [12]) for the use of Simulcrypt in DVB systems.

11.5.2 Typical Simulcrypt implementation

In the implementation described here, Simulcrypt is used in such a way that it results in the sharing of a transponder channel, so that the same program or the total programming package can be received simultaneously by the set-top box populations of two service providers in different geographical areas. Figure 11.3 shows a functional description of the technical operation of Simulcrypt.

In Figure 11.3, there are two possibilities for receiving. The first is via a subscribers' satellite antenna and the second is via a CATV operator's cable head-end. Figure 11.3 uses set-top boxes of service provider A and respectively service provider B. Provider B adds its ECMs and EMMs to the MPEG-2 TS at the uplink by using a *control word* (CW) provided by service provider A. The CW provides access to provider A's encoder/multiplexer.

This requires a common framework for the indicating of the various ECMs and EMMs data flows and a common encryption system. Both conditions have been met—first by the specifications of the MPEG-2 TS and second by the DVB members' work, which resulted in the CSA.

When both providers' ECMs and the EMMs are sent along, both set-top box populations can receive the same program simultaneously and the expensive transponder costs can be shared. Also, this implies the use of but a single uplink in Europe (this could also apply to the United States or Japan), which uses the CSA and supports proprietary CAMSs.

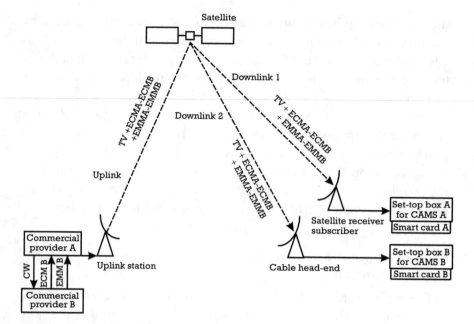

Figure 11.3 Functional description of Simulcrypt. (*Source:* DVB project.)

Furthermore, the subscriber only needs one set-top box. Finally, there is now an impulse for (further) investments. The CA service providers' previous investments in the set-top boxes remain protected, because by applying Simulcrypt one's own CAMS can be used.

11.6 Transcontrol

The large investments that are inherent to the provision of CA services, have deterred CATV operators from entering this turbulent market. The vertically integrated entities that did make these investments simply rented their transmission capacity. Today, CATV operators increasingly want to extent their service provision with CA services. Moreover, they want to regain the control over the management of the services that are provided via their own networks. DVB decided to support the use of *Transcontrol*, which allows CATV operators to have full control over the services at a local level. This section first explains the Transcontrol model and subsequently discusses a technical implementation of Transcontrol at a functional level.

11.6.1 The Transcontrol model

As stated earlier, the traditional vertically integrated service providers control the complete CA system and rent television channels from satellite and cable-TV operators. The set-top boxes are either sold or rented to the subscribers. CATV operators are striving to add value to their CATV networks. In the new constellation, they want to provide value-added services for which subscribers have to pay additional fees. This contrasts with the provision of transmission capacity only, as in the past. The means by which this can be achieved is a CA system. With now two competing CA service providers, this implies that a subscriber would have to rent or buy two different set-top boxes. If the management of the CA system could be handled by one provider, only one set-top box would be required.

After (again) intensive debate, the members of the DVB project have agreed that the CATV operators must be able to have complete control on a local or regional level over the services that use digital CA systems. In technical terms this implies that every CATV operator must be at liberty to replace the CAMS with its own CAMS if and when the operator wishes to do so. DVB does not clarify whether the control over the CAMS concerns the SMS and SAS or the SAS only. The latter seems a feasible solution, as the SMS typically includes proprietary information. The handover of the control over the services is referred to as Transcontrol.

The principle of Transcontrol is included in the Directive on Television Standards but is limited to the case of CATV operators, rather than being applicable to all network service providers that directly deliver television programs to consumers via their own CA systems. In the present digital era, all different kinds of multimedia services can be provided via various networks. By the application of compression techniques video signals can, for example, be provided via PSTNs. Hence, it is feasible to apply Transcontrol as a more general principle.

11.6.2 Typical Transcontrol implementation

With the application of Transcontrol, the CA service provider's CAMS is replaced by a CATV operator's own CAMS. Figure 11.4 shows an example of a satellite operator sending the signals of three service providers through to the cable head-ends of, in this case, three different CATV operators.

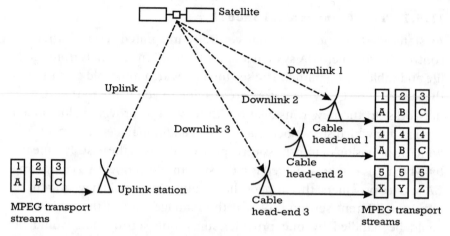

Figure 11.4   Functional description of Transcontrol.

In this example, the MPEG-2 TSs from the three service providers are presented to an uplink station. A service provider's TS consists of a CSA-encrypted program and a proprietary CAMS. The CSA encrypted programs are shown with the letters A, B, and C. The corresponding CAMSs are labeled 1 to 3, respectively.

The TSs are then sent to the satellite via the uplink. The satellite sends the streams to the cable head-ends unchanged. In the case of downlink 1, the TSs are also sent unchanged to the subscribers by the CATV operators. In this case, the subscriber needs three different set-top boxes. In the case of downlink 2, the CAMS is replaced by the CATV operator's own CAMS, indicated by the number 4. Thus, in this situation, Transcontrol is used and one set-top box is sufficient.

In addition to replacing the CAMS, another solution is to decrypt the encrypted program and to encrypt it again. Take, for example, downlink 3. Some consider this to be another form of Transcontrol. This is, however, a broad interpretation because beside the replacement of the CAMS, the program too is re-encrypted without this adding anything to the functionality of Transcontrol. Moreover, the question is whether it is desirable to incorporate this possibility in the CA system. After all, it provides a possibility for a compromise of the systems' security.

Figure 11.4 presents a typical implementation of Transcontrol at a functional level. Figure 11.5 depicts a possible technical implementation of Transcontrol.

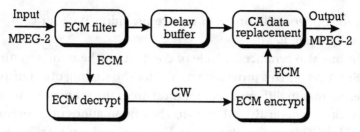

Figure 11.5 ECM regeneration. (*Source:* Irdeto Consultants.)

This scheme shows that the ECM is filtered from the MPEG-2 TS. Next, the concerned CW is regained by the decryption of this ECM. The CW is encrypted again with the proprietary crypto system of the CATV operator, which results in a new ECM. By means of a *drop-and-add* multiplexer the data related to the CAMS (i.e., the CAMS data for letting an authorized smart card decrypt the program) is replaced by the data of the new CAMS and is inserted in the delayed MPEG-2 TS. Because of the delay as a result of receiving and the regeneration of the ECM, the MPEG-2 TS in turn is delayed for about 1 second.

11.7 Summary and conclusions

Because of the different interests of the various parties involved, CA has been an area of intensive debate in DVB. A balance had to be struck between opening up protected markets and at the same time not undermining the investments of the current, often vertically integrated, service providers. This has, among other things, resulted in the standardization of a common framework for the encryption of television programs, namely the CSA.

It was decided not to standardize the CAMS. An important reason was that the current service providers had already made big investments in their proprietary CAMSs, particularly in their SMS including the client data. Hence, a standardized interface between the proprietary CAMSs and the programs encrypted with the CSA was needed. For this purpose DVB standardized the CI and, by doing so, supported the Multicrypt model. In this respect, the CI can be regarded as a solution to open up the market for horizontally oriented service providers (i.e., for competitors on a level playing field).

DVB has also produced a code of conduct for the nondiscriminatory use of Simulcrypt. This provides a model for the sharing of a transponder channel, so that in different geographical areas the same program can be received simultaneously by the set-top box populations of different service providers. This implies that vertically integrated service providers can no longer protect their market by excluding others from making use of their proprietary CA system.

Moreover, DVB decided to support the use of Transcontrol. With Transcontrol, CATV operators are able to control the services that use digital CA systems at a local or regional level. The use of Transcontrol is limited to CATV operators only, rather than being applicable to all network service providers that directly provide television programs to consumers via their own CA systems. Table 11.1 provides an overview of situations in which the different CA models are applied and indicates the parties to which this is particularly beneficial. Depending on the situation, combinations of these models may be used.

The European Commission, in cooperation with several Member States, has supported the work of DVB by constructing the Directive on Television Standards. This Directive, among other things, sets the standards for the CSA mandatory. The CI is mandatory for television sets with a built-in host but is not mandatory for use in set-top boxes. Hence, Multicrypt is supported in a "limited" way. Moreover, the directive supports the Simulcrypt model by complying with the code of conduct. Finally, the

Table 11.1
Application of CA Models

CA Model	Application	Particular Benefit
Multicrypt	Many horizontally oriented service providers provide their services to one and the same set-top box	Horizontally oriented service providers and the user
Simulcrypt	The same transponder is shared by two or more different vertically oriented service providers that operate in different geographical areas; the user needs one set-top box only	Vertically oriented service providers and the user
Transcontrol	Many service providers provide their services via a CATV network to one and the same set-top box	CATV operators and the user

Transcontrol model is adopted in the Directive **but is** limited to the case of CATV operators. As DTV programs can be provided via all different kinds of networks to the end user (e.g., a PSTN (*public switched television network*) or a terrestrial network), it would be feasible to apply Transcontrol as a more general principle. However, this has not been decided.

References

[1] OECD Working Party on Telecommunication and Information Services Policies, *Conditional Access Systems: Implications for Access*, DSTI/ICCP/TISP(97)7, Paris, September 15–16, 1997.

[2] Wiemans, F. P. E., J. M. Smits, H. C. A. van Tilborg, *Encrypie: justitiële en particuliere belangen; Een verkennende beschouwing*, in: Delict en delinquent Nr. 24, April, 1994, p. 341.

[3] Diffie, W., and M. E. Hellman, "New Directions in Cryptography," *Trans. IEEE on Information Theory*, IT-22, No. 6, November, 1976, pp. 644–654.

[4] Rivest, R. L., A. Shamir, and L. Adleman, *A method of obtaining digital signatures and public key crypto systems*, Comm. ACM, 21, No. 2, February, 1978, pp. 12–126.

[5] van der Lubbe, J. C. A., *Basismethoden voor cryptografie*, Delftse Uitgevers Maatschappij, 1994, p. 191.

[6] de Bruin, R., "The Key to Information Security," *Telecommunications International*, January 1997, pp. 55–57.

[7] de Bruin, R., "Making Interactive TV Pay," *Telecommunications International*, September, 1997, pp. 105–108.

[8] EBU/CENELEC/ETSI-JTC, "Digital Video Broadcasting (DVB); Support for use of scrambling and Conditional Access (CA) within digital broadcasting systems", ETR 289, October,1996.

[9] Directive 95/47/EC of the European Parliament and of the Council of 24 October 1995 on the use of standards for the transmission of television signals, O.J. L281/51, 23 November, 1995.

[10] DVB, *DVB Common Scrambling Algorithm; Distribution Agreements*, DVB Document A011, rev. 1, June, 1996.

[11] DVB, *Common Interface Specification for Conditional Access and other Digital Video Broadcasting Decoder Applications*, DVB Document A017, May, 1996.

[12] DVB, *Technical Specification of Simulcrypt in DVB Systems; Part 1: Head-end and Synchronization*, TS101 197-1, June, 1997.

Contents

Interactive services

12.1   Introduction

The basic principle of services such as television or pay-TV is that the content is distributed via a broadcast network to the end user. When these types of services are taken into account, television can be considered a passive medium. Interactive television implies that in communication the end user is able to control and influence the subjects of communication, with the control and influence taking place via an interaction network. Examples of interactive services are Internet services that are provided via the television medium, video mail, and interactive teletext. If interactive services are provided on the basis of CA, one can, for example, think of pay-per-view or video-on-demand.

Before developing technical standards for interactive services, DVB first specified the commercial requirements [1]. One of these requirements is that the specifications must be compatible with different types of networks. DVB decided not to specify an interaction channel solution associated with each broadcast system. Hence, in the higher layers DVB developed network independent protocols for interactive services [2, 3]. However, at the transport and physical level, several standards have been specified for the different networks. One of these standards concerns broadcasting via CATV networks, where the interaction channel(s) can be embedded in the CATV network itself. In case of the DVB satellite and terrestrial broadcasting systems, however, it is specified that the interaction channels are accommodated in a separate network. The application layer, as well as the hardware and software of the end user terminal, are left up to the market.

The DVB interactive services model is based on the use of a set-top box. However, the use of a set-top box is not required for all types of interactive services, especially those services that are available at no charge. Therefore, this chapter also discusses the provision of Internet services and interactive teletext via the television medium, without the application of set-top boxes.

12.2   Elements of interactive services

Several (technical) elements play a role in the provision of interactive services. This section discusses the reasons for (not) using interaction channels (i.e., providing interactive services) in the first place; explains the typical transmission characteristics, which have to be regarded when specifying an interactive communication system, and describes a generic interactive systems model.

12.2.1   Reasons for (not) using interaction channels

Interactive television requires at least a return interaction path from the end user to the interactive service provider. This is referred to as unidirectional interaction. Bidirectional interaction requires an additional forward interaction path from the interactive service provider to the user. As it concerns control data only, both paths are narrowband channels. The control data sent via the return interaction path consists of application control data or application communication data. Additionally, the forward interaction path may carry data download control information.

There are a number of reasons for using interaction channels [4]. First, enhanced security can be achieved, because a one-to-one link between the user and the interactive service provider can be established. However, communication via open networks can be intercepted. Hence, communication via interaction channels should be encrypted. Second, payment billing can be achieved in a more cost-efficient and less time-consuming way. For example, (impulse) pay-per-view and video-on-demand programs can be registered automatically by using a return interaction path, rather than the user having to make a telephone call to the interactive service provider. Third, a return interaction path could be used to collect diagnostic information related to the transmission quality such as signal strength or BER or other statistical information concerning the programs watched. Finally, in case of large shared networks, the capacity for the transmission of entitlement messages may be insufficient. A forward interaction path can be used to achieve additional capacity. Moreover, a return interaction path can be used to check that the interactive set-top box is tuned to the correct channel when sending entitlement messages. This could reduce the number of messages that have to be repeated perpetually.

There are also a number of reasons for not using interaction channels. First, the set-top box costs increase. Next, installation difficulties may be introduced when using a telephone line as an interaction channel. A user may not have a telephone in the relevant room. In this case an extension cord or a cordless connection is required, which also leads to extra costs. Finally, the use of a telephone line for the establishment of an interaction channel makes it impossible to make or receive normal phone calls (or other terminal equipment cannot make use of the telephone line) when the set-top box is communicating, unless there is another telephone in the house.

The use of interaction channels introduces significant benefits and can be cost-effective. It would be preferable that set-top boxes, which are not manufactured for interactive services, would have the capability to implement at least a return interaction path. However, this is for the market to decide.

12.2.2 Out-of-band/in-band signaling

Interactive systems utilize at least a return interaction path, and an additional forward interaction path may be used for downstream signaling. The forward interaction path can be established in two different ways. Under the first option, it can be established by means of a forward interaction path via a separate interaction network. This path is reserved for interactivity and data and control information only. This option is called *out-of-band* (OOB) downstream signaling.

Alternatively, the forward interaction path can be incorporated in the broadcast signal. This option is referred to as *in-band* (IB) downstream signaling. In the case of DVB, IB downstream signaling implies that the forward interaction path is embedded in the MPEG-2 TS.

The application of either OOB or IB downstream signaling implies different requirements for set-top boxes. Depending on the type of downstream signaling used, one can speak either of an OOB set-top box or of an IB set-top box. However, both systems may coexist on the same network on the condition that different frequencies are used for each system.

12.2.3 Spectrum allocation

The processing of the return interaction path and the forward interaction path via one and the same network requires spectrum allocation. To avoid any interference, the frequencies of the return interaction path and the forward interaction path need to be allocated in a different frequency range. A sufficient guard band between these frequency ranges should be respected to avoid filtering problems in the bidirectional video amplifiers and in the set-top boxes (see Figure 12.1).

12.2.4 Multiple access techniques

The *frequency division multiple access* (FDMA) technique is used for transmitting several message signals over a communication channel by dividing the available bandwidths into slots, one slot for each message signal [5]. A small guard band between the slots is used to avoid interference between adjacent channels.

The *time division multiple access* (TDMA) technique divides a time frame into slots. Each slot is a time period during which a message can be transmitted over the communication channel. The full channel bandwidth is

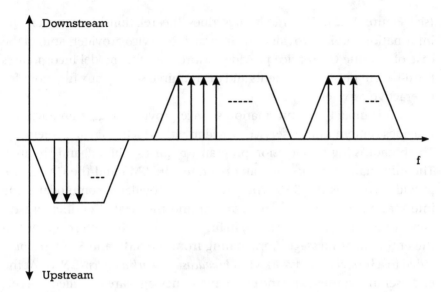

Figure 12.1 Frequency allocation of interaction channels.

available to transmit traffic bursts in each time slot. TDMA allows the transmit start times to be synchronized to a common clock source. Synchronizing to a common clock increases the message throughput of the communication channel.

In principle, FDMA and TDMA accomplish the same results. However, nonlinearities in the circuits of the FDMA system can result in intermodulation and harmonic distortion, which lead to interference between adjacent channels. This affects both high-frequency and low-frequency channels. In case a large number of channels is being multiplexed, the requirements for the FDMA systems' circuits become very stringent. Because in TDMA the message signals from the different channels are not processed simultaneously but sequentially, these requirements do not apply. Moreover, the digital TDMA circuits are simpler, more reliable, and efficient in operation. Hence, the TDMA is often applied as multiple access technique.

12.2.5 Generic interactive systems model

For the purpose of explaining the operation of several interactive television services, a generic model for interactive television systems is used (see Figure 12.2). This model describes the relationship between the information service provider, the interactive service provider, and, in the case of CA, the CA service provider. Moreover, this model incorporates the different network elements and the required set-top box functions for interactive services.

In this model, the information service provider (e.g., a broadcaster) provides content to a CA service provider with which it has an agreement. The broadcasting of television programs requires a broadband channel. The information service provider operates the SMS and the CA service provider controls the SAS. The CA service provider accommodates the information service providers' content and the related entitlements to view (originating from the SMS) in his CA-system. Next, the content and the entitlement messages (originating from the SMS and SAS) are provided to a broadcast network via a *broadcast network adapter* (BNA). At the end user, the content and the entitlement messages are provided to a set-top box. This set-top box incorporates a *network interface unit* (NIU), which in turn consists of a *broadband network interface* (BNI) and a *interactive network interface* (INI), and a *set-top-unit* (STU). The interactive service provider can provide its services via the information service provider or to the BNA directly. In case of CA, the interactive service provider can either provide its services to the information service provider or via the CA provider directly. In the latter case, the interactive service provider controls the SMS.

The model shows that the set-top box incorporates an INI. However, this interface may also be a module external to the set-top box. The return interaction path and the OOB forward interaction path are (narrowband) channels within the interaction network. The connection between the interaction network and the interactive service provider is realized via an *interactive network adapter* (INA). Alternatively, an IB forward interaction path, rather than an OOB forward interaction path, may be used.

12.3   DVB interaction channel for CATV networks

This section explains how several basic elements of interactive services, which were discussed in the previous section, are respected by DVB in the specifications concerning interaction channels for CATV networks.

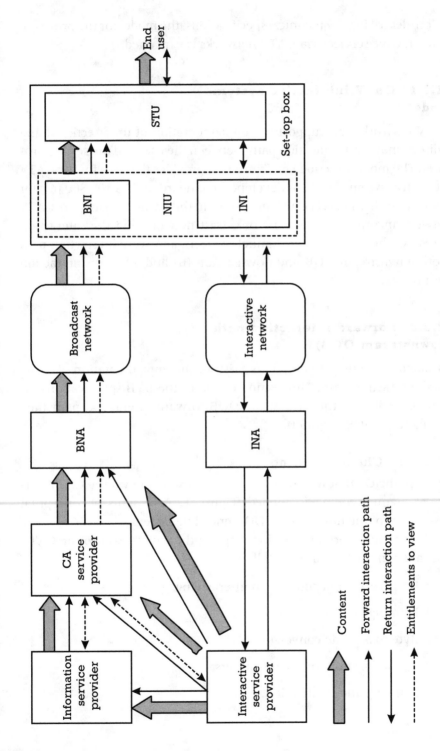

Figure 12.2 Generic model for interactive television systems.

Before describing the technical specifications, the model for the provision of interactive services via CATV networks is explained.

12.3.1 CATV interactive system model

CATV networks can support the implementation of unidirectional and bidirectional communication paths between the user and the service provider. The model described in Figure 12.2 provides the basis for the CATV interactive system. DVB has developed a standard (prETS 300 800 [6]) for interaction channels via CATV networks. In this particular case, the CATV network incorporates both the broadcast network and the interaction network. These interaction channels concern a forward interaction path (both downstream OOB and downstream IB) and a return interaction path (upstream).

12.3.2 Forward interaction path (downstream OOB)

As stated previously, two options exist for the implementation of a forward interaction path. This section describes the DVB specifications for channel coding in the case of the OOB forward interaction path (i.e., downstream OOB channel).

12.3.2.1 Channel coding

To adapt the OOB downstream signal to transmission via the CATV network, channel coding is applied. Figure 12.3 presents a conceptual description of the downstream OOB encoding system.

The most important steps for adapting the data stream to the CATV transmission medium are the following:

▶ Error correction coding and interleaving;

▶ Framing;

▶ Byte to m-tuple conversion;

▶ Randomization for energy dispersal;

▶ Mapping and modulation.

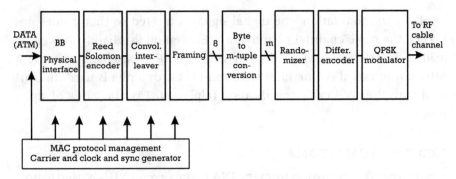

Figure 12.3 Conceptual downstream OOB encoding system description.

Several building blocks of this system description are discussed in Chapter 10. Hence, only the different parameters of these elements and the new elements are discussed. Moreover, with the help of the recovered carrier and clock signals and sync signal, the decoding system more or less reverses the encoding process at the receiving end. Therefore, the decoding system is not discussed.

12.3.2.2 Spectrum allocation and filtering

In general, the exact location of the frequency band in which CATV systems operate, is chosen by the CATV network operator. However, to simplify the NIU's tuner, DVB has provided a guideline for the use of frequency ranges. Hence, it is preferable to use the 70–130 MHz and/or 300–862 MHz frequency bands, or parts thereof, for the OOB forward interaction path. Table 12.1 presents the channel bandwidths that are used for the forward interaction channel.

Table 12.1
Channel Bandwidths for the OOB Channels

Grade	Channel Bandwidth
Grade A	1 MHz
Grade B	2 MHz

Prior to modulation, the digital signal is filtered so that it does not exceed the cable channel's bandwidth. Exceeding this bandwidth could lead to interference with adjacent channels. For baseband shaping a square root raised cosine filter with roll-off factor alpha is used. In contrast with the DVB cable specification (alpha = 0.15), the value of alpha is 0.30.

12.3.2.3 FDMA/TDMA

Downstream transmission from the INA to the several NIUs is used to provide synchronization and information to all set-top boxes. This allows the NIUs to adapt to the network and send synchronized information upstream. DVB has specified a multiple access scheme in which an address is assigned to each user. A *media access control* (MAC) address is stored in the set-top box for user identification. As such, it represents the NIU's unique MAC address. This 48-bit address may be hard coded in the NIU or provided by an external source. In case different CATV networks are involved, an additional network address is required.

With the utilization of the TDMA technique by DVB, each upstream channel is shared by many different users. These upstream channels are all divided into time slots that can be accessed by the users. The packets can either be sent by the user with a possibility of collision, or they can be transmitted during a time slot that is assigned by the INA. In case message slots are not in use, an NIU may be assigned multiple message slots for increased messaging throughput. The additional message slot assignments are provided via the downstream channel.

Each downstream channel contains a synchronization frame, by which synchronization of up to eight upstream channels can be achieved. The frequencies of the upstream channels are indicated by the MAC protocol. In order for all NIUs to work with the same clock, a time reference at the INA is sent periodically via the forward interaction path and received simultaneously by all NIUs. Hence, the slot times for all NIUs are aligned.

The following access modes for the upstream slots are specified:

▶ *Contention access:* Contention access is used for multiple users that have equal access to the upstream signaling channel. It can be used to send either MAC messages or data. In general, the OOB MAC messages consist of 40 bytes, but they may be longer. In case of simultaneous transmission, collision may occur. A collision occurs when two or more NIUs attempt to transmit a packet during the same time slot on the same channel. A contention resolution protocol is used to solve this problem. For each packet transmitted, the NIU has to receive a positive acknowledgment from the INA. This positive acknowledgment implies that a collision did not take place. If it did (i.e., the NIU did not receive a positive acknowledgment), the NIU retransmits the packet concerned.

▶ *Fixed-rate access:* The user has a reservation of one or more time slots in each frame. The INA uniquely assigns a slot to a connection. The NIU cannot initiate a fixed-rate access.

▶ *Reservation access:* To satisfy the users' need for more transmission capacity, the NIU sends a request for more time slots to the INA than have been reserved initially. The INA uniquely assigns these slots to a connection on a frame-by-frame basis.

▶ *Ranging access:* It may occur that a slot is preceded and followed by slots that are not used by other users. By means of these upstream slots, the time delay and power can be measured and adjusted.

The different access modes may be used on a single carrier. This enables different services on one carrier only. On the other hand, a carrier can also be assigned to one specific service. In this case, only the slot types that are needed for this service will be used. This allows the terminal to be simplified.

12.3.2.4 Randomizing

After the byte-to-bit mapping process, the data stream is applied to a randomizer to ensure a pseudo random distribution of ones and zeroes. The randomizer constitutes a *linear feedback shift register* (LFSR), which is presented in Figure 12.4.

An initial sequence is loaded into the LFSR. The serial output results from the serial input and the feedback loop. The latter is constructed by an AND-operation of the first bit (i.e., MSB) at the output of the LSFR and the bit that follows the MSB. At the receiving end, a complementary self-synchronizing derandomizer is used to recover the data.

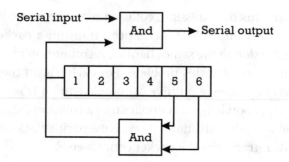

Figure 12.4   Randomizer.

12.3.2.5 Bit Rates and framing

For the forward interaction path, several grades (i.e., bit rates) can be used. The implementation of only one of these bit rates is mandatory. Table 12.2 presents an overview.

DVB has specified an OOB signaling frame format. A frame consists of 193 bits. One bit is reserved for overhead, and the payload consists of 192 bits (24 bytes). An extended superframe is constructed by 24 of these frames. Hence, the extended superframe consists of 4,632 bits (see Figure 12.5).

Each frame's overhead accommodates one information bit for synchronization of upstream slots. Together, the 24-frame overhead bits are divided into six bits for extended superframe alignment, six bits for cyclic redundancy check bits, and 12 data link bits. The extended superframe alignment is used to locate all 24 frames and overhead bit locations. The cyclic redundancy check bits allow a redundancy check between sequential extended superframes. The M-bits serve for slot timing assignment.

The extended superframe's payload structure accommodates combinations of an ATM cell (53 bytes) and corresponding RS parity values (2 bytes). According to the ITU standard [7], the ATM cell format, in turn,

Table 12.2
Bit Rates for OOB Channels

Grade	Bit Rate
Grade A	1.544 Mbps
Grade B	3.088 Mbps

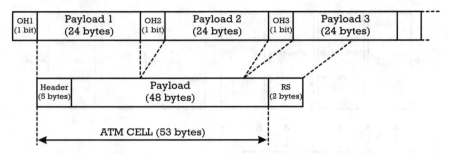

Figure 12.5   OOB signaling frame format.

consists of a 5-byte header and a 48-byte payload. The RS encoding performs the correction of one erroneous byte per ATM cell.

12.3.3   Forward interaction path (downstream IB)

Section 12.3.2 describes the DVB specification for the OOB forward interaction path. The alternative is to implement an IB forward interaction path (or downstream IB channel); this section discusses the DVB specifications for this method.

12.3.3.1   Channel coding

The IB forward interaction path is embedded in the broadcast signal. In case of the DVB specification, this implies that the information concerned is incorporated in the MPEG-2 TS. The specifications for the IB forward interaction path comply with the DVB cable standard (ETS 300 429), which is discussed in Chapter 10 (note that baseband shaping is not applied). Figure 12.6 presents a conceptual representation of the encoding system and modulation. The elements, which are common with the DVB cable standard, are presented in gray.

As the decoding process by means of the recovered carrier and clock signals and sync signal more or less reverses the encoding process, the downstream IB decoding system is not discussed.

12.3.3.2   Bit Rates and framing

For the IB forward interaction channel no other constraints exist than those that are specified in the DVB cable specifications (see Chapter 10).

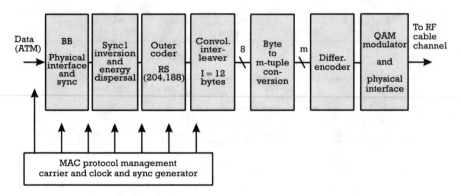

Figure 12.6 Conceptual downstream IB encoding system description.

However, DVB produced a guideline to use bit rate multiples of 8 Kbps. MPEG-2 TS packets with a specific PID are sent at least in every period of 3 ms to achieve synchronization of upstream slots.

Figure 12.7 presents the frame structure of the IB forward interaction channel, which carries MPEG-2 TS packets.

The 4-byte MPEG header contains a specific PID assigned to MAC messages. The upstream marker field, which consists of 3 bytes, provides upstream synchronization information. The next field accommodates a 2-byte slot number. The 3-byte MAC flag control field provides control information, which is used in conjunction with the MAC flags and extension flags. The MAC flag field, which consists of 26 bytes, contains eight slot configuration fields of 3 bytes each. These fields, in turn, contain slot configuration information for the related upstream channels and are followed by two reserved bytes. The 26-byte extension flags field is used in case one or more 3.088 Mbps upstream channels are used. The definition of this field is identical to that of the MAC Flag field. Next, three MAC message fields accommodate a 40-byte MAC message each. The general format for an IB MAC message is limited to 120 bytes. Finally, a 4-byte field is reserved for future use.

MPEG header (4 bytes)	Upstream marker (3 bytes)	Slot number (2 bytes)	MAC flag control (3 bytes)	MAC flags (26 bytes)	Extension flags (26 bytes)	MAC message (40 bytes)	MAC message (40 bytes)	MAC message (40 bytes)	Res. for future use (4 bytes)

Figure 12.7 IB signaling frame format (MPEG-2 transport stream format).

12.3.4 Return interaction path (upstream)

The DVB specifications for the OOB and IB forward interaction paths were described in Sections 12.3.2 and 12.3.3. A return interaction path (or upstream channel) is necessary to enable the user to send information to the interactive service provider. This section discusses the DVB specifications for the return interaction path.

12.3.4.1 Channel coding

Channel coding is applied to adapt the return signal to transmission via the CATV network. A conceptual description of the upstream encoding system is presented in Figure 12.8.

The most important steps for adapting the data stream to transmission via a CATV network are the following:

▶ Error correction coding;

▶ Byte to *m*-tuple conversion;

▶ Randomization for energy dispersal;

▶ Mapping for modulation;

▶ Addition of unique word;

▶ Modulation.

Chapter 10 discusses several elements of the encoding system. Accordingly, this chapter discusses only the different parameters of these elements and the new elements. By means of the recovered carrier and

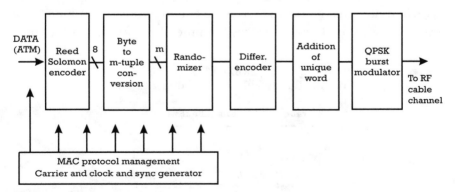

Figure 12.8 Conceptual upstream encoding system description.

clock signals and the sync signal, the decoding system more or less reverses the encoding process at the receiving end. Hence, the decoding system is not discussed.

12.3.4.2 Spectrum allocation and filtering

To simplify the tuner of the NIU, DVB has provided a guideline for the use of frequency ranges. The 5–65 MHz frequency band, or parts thereof, is preferred for use with the return interaction path. Table 12.3 presents the channel bandwidths.

Prior to modulation, the digital signal is filtered so that it does not exceed the cable channel's bandwidth. Exceeding this bandwidth could lead to interference with adjacent channels. For baseband shaping a square root raised cosine filter with roll-off factor alpha is used. In contrast with the DVB cable specification (alpha = 0.15), the value of alpha is 0.30.

12.3.4.3 Unique word

The different users send the upstream packets independently. Hence, the TDMA traffic consists of a set of bursts originating from a number of users. The TDMA system periodically transmits one or more upstream bursts within time frames. Each frame normally consists of two (primary and secondary) reference bursts that are used for timing reference, traffic bursts that carry digital information, and the guard time between bursts to avoid interference in adjacent channels (see Figure 12.9) [8].

A user accessing a channel may transmit one or more traffic bursts per frame and may position the traffic burst(s) anywhere in the frame. The reference burst accommodates a unique word, which provides the receive timing that allows the receiving end to locate the position of the traffic burst in the frame. It marks the time of occurrence of the traffic

Table 12.3
Channel Bandwidths of the Return Channels

Grade	Channel Bandwidth
Grade A	200 kHz
Grade B	1 MHz
Grade C	2 MHz

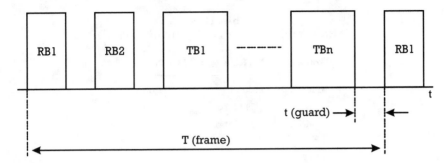

Figure 12.9 TDMA frame structure.

burst and provides the receive burst timing that allows the receiving end to subtract only the required sub-bursts within the traffic burst.

To enhance detection, the unique word is a sequence of ones and zeroes with good correlation properties. The unique word detection is used to mark the receive frame timing if the unique word belongs to the primary reference burst. It marks the receive traffic burst timing if the unique word belongs to the traffic burst. Hence, the position of every burst in the frame is defined with respect to the receive frame timing, and the position of every sub-burst in the traffic is defined with respect to the burst's receive burst timing.

Accurate detection of the unique word is of utmost importance in a TDMA system. The entire traffic burst is lost when a traffic burst's unique word is missing. This implies that the BER has increased. A false detection of the primary reference burst unique word generates the wrong receive frame timing and consequently incorrect transmit frame timing. This causes out-of-synchronization transmission, which results in overlapping with other bursts.

12.3.4.4 Bit rates and framing

Three grades (i.e., bit rates) can be used for transmission via the return interaction path. Only the implementation of one of these bit rates is mandatory. The upstream packets, which are sent in a bursty mode from the different users, consist of 512 bits each. Hence, a bit rate of 256 Kbps corresponds to a slot rate of 500 slots/s. Table 12.4 presents an overview.

The upstream information either concerns data or MAC messages. DVB specified that MAC messages, which are sent via the return

Table 12.4
Bit Rates and Slot Rates of Upstream Channels

Grade	Bit Rate	Slot Rate
Grade A	256 Kbps	500 slots/s
Grade B	1.544 Mbps	3000 slots/s
Grade C	3.088 Mbps	6000 slots/s

interaction channel, are limited to 40 bytes. Additionally, DVB specified a frame format for the return interaction path (see Figure 12.10).

A burst mode acquisition method is provided by a 4-byte unique word. The 53-byte payload area contains a single message cell. The format of the message cell is consistent with the ITU standard for ATM cells. Next, a 6-byte RS parity field achieves the correction of 3 erroneous bytes over the payload area. Finally, 1 byte is used as a guard band, which provides spacing between adjacent channels.

12.4 DVB interaction channel through PSTN/ISDN

This section discusses the DVB baseline specifications for both unidirectional and bidirectional interaction paths via PSTNs and ISDNs. First, the interactive system model for PSTN/ISDN is described. Next, the technical

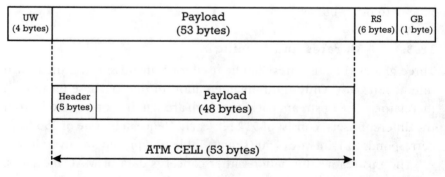

Figure 12.10   Return channel frame format.

specifications for the provision of interactive services by means of a bidirectional interaction path through PSTN are discussed. Finally, the technical specifications for the bidirectional interaction path via ISDN are explained.

12.4.1 PSTN/ISDN interactive system model

In principle, PSTN/ISDN can support the implementation of both unidirectional and bidirectional interaction paths from the user to the interactive service provider. The model described in Figure 12.2 provides the basis for the PSTN/ISDN interactive system. In this case the interaction network is formed by either a PSTN or an ISDN. The broadcast network can be a DVB system for broadcasting via a satellite channel, a CATV network, or a terrestrial network.

DVB has developed specifications (prETS 300 801 [9]) for an interaction channel through PSTN/ISDN. These interaction channels concern a forward (downstream) and a return (upstream) interaction path.

12.4.2 Interaction path through PSTN

A PSTN is an analog circuit-switched network that provides narrowband bidirectional channels for analog transmission with a bandwidth of 4 kHz each. For the digital DVB broadcasting systems, one of these channels can be used to implement the return interaction path. A modem is required for bidirectional communication between the user and the interactive service provider. This modem constitutes a user interface module to the network and can either be internal or external to the set-top box. The interface between the modem and the PSTN must meet the existing ETSI requirements [10] for a logical interface and the concerned DVB physical interface requirements [11].

Because the interaction takes place via a public telecommunications network (i.e., the PSTN), the modem is considered terminal equipment. Within the EU this has legal implications. The interface between the modem and the PSTN has to comply with the existing legal requirements [12] for terminal equipment. If the modem is internal to the set-top box, the whole set-top box is considered to be terminal equipment. This contrasts with a set-top box that is connected to a broadcast network and is used for receiving broadcast signals only. In case a CATV network is considered to be a public telecommunications network, the same legal requirements are applicable to interactive set-top boxes that are connected to this network.

As stated in Section 12.2.1, the use of a telephone line for the establishment of an interaction channel makes it impossible to make or receive normal phone calls when the set-top box is communicating, unless there is another telephone line available. While the modem is in use, other terminal equipment cannot make use of the telephone line either. In some countries users do not even have the capability to interrupt an active communication that is established via the PSTN. To enable emergency calls, DVB specified that during dialing or data transfer the modem connection can be cut off. Additionally, the modem performs a forced disconnection in case the user hooks off any other terminals, which are connected via the same telephone line.

When the modem is the called party, it may not be possible to interrupt the communication. For this case, DVB specified that the modem does not accept any incoming calls from any interactive service provider. Hence, the modem always initiates the call to the interactive service provider to establish a bidirectional return channel. Alternatively, disconnection can be implemented in higher layer protocols.

12.4.3   Interaction path through ISDN

An ISDN is a digital circuit-switched network that provides two 64-Kbps channels (B channels) for data transfer and one signaling channel (D channel) with a bit rate of 16 Kbps. Nowadays, ISDN is commonly used in public telecommunications networks.

This type of network can either be used to implement unidirectional or bidirectional interaction paths for the provision of interactive services. Basic rate access (64 Kbps) can be applied in case of bidirectional interaction between the user and the interactive service provider. The interface between the set-top box and the ISDN must meet the existing ETSI requirements [13] for a physical interface, as well as the current ITU/ETSI logical interface requirements [14, 15].

In case interaction between the user and the interactive service provider takes place via a public ISDN, the set-top box is considered terminal equipment. Hence, the interface between the set-top box and the ISDN has to comply with the existing legal requirements for terminal equipment. If the ISDN concerned is a private network, these legal requirements are not applicable.

Because the ISDN supports two channels for data transfer, it is possible to make or receive normal phone calls when the set-top box is communicating. When other terminal equipment is connected to the other channel, incoming and outgoing phone calls can no longer be made. The signaling channel can be used by higher layer protocols to disconnect in case of emergency calls.

12.5   Internet services via broadcast networks

Beside the work of DVB, which is mainly focused on the use of a set-top box, there are other interesting developments in the field of interactive services. This section discusses an alternative approach for the provision of interactive services via broadcast networks, for which a set-top box is not required. This particular case concerns interactive Internet services. Before going into detail, a short introduction to the background and history of the Internet is first presented.

12.5.1   The Internet

In 1973 the *Defense Advanced Research Projects Agency* (DARPA) in the United States started a project to develop a technology with which an internet (i.e., a series of networks) could be created. For this purpose, the *transmission control protocol/Internet protocol* (TCP/IP) was developed. The first network to make use of this technology was called ARPANET (the "D" of DARPA was left out). In particular, research institutes that were under the supervision of the Department of Defense obtained access to this network.

In 1980, the ARPANET was divided into two parts, the MILNET for military applications and a part that continued to exist under the name ARPANET. The National Science Foundation decided in 1985–86 to build a national network, based on the TCP/IP technology. This network was called the NSFnet and became the infrastructure to which, in principle, all universities and research institutes could be connected. At the end of the 1980s, the NSFnet developed branches to other parts of the world.

Despite the fact that in the mid 1980s a global network for electronic mail services already existed (EARN/BITNET), the NSFnet became popular so fast that many universities and research institutes around the world switched to TCP/IP-based networks. This led to the current Internet, which consists of a series of international, national, and local networks. What these networks have in common is that they all use the same protocol (TCP/IP).

12.5.2 Internet services via CATV networks

The Internet provides a global VAN that is built up of different types of networks. The backbone is formed by networks with a large transmission capacity (e.g., ATM or SDH networks). The user is often connected via a local loop network that has been designed for telephony services (e.g., a PSTN) for which a limited transmission capacity is required. Internet users can access all different kinds of multimedia services, among other information, and download files from Internet sites. The use of these services requires more bandwidth than the traditional telephony services. Internet access via these low-capacity networks often results in congestion. The user may notice, for example, that when accessing an Internet site, some time is needed to build up the pictures on the screen. Moreover, even when a high-speed modem is used, the downloading of files can sometimes take over an hour. Moreover, the congestion increases considerably when a large number of users are connected to the Internet at the same time.

Additionally, the users are often connected to broadcast networks that have been designed for broadcasting high-bandwidth information (i.e., video signals). Although these networks were not originally designed for point-to-point communication, they can be used to provide fast Internet access. For, example, a CATV network can be used to transmit broadband information to a specific user, while the PSTN provides the required return channel. Hence, the channel bandwidth increases dramatically, and real-time access is possible. Different systems can be constructed to provide Internet services via broadcast networks. As an example, Figure 12.11 presents a system for the provision of fast Internet services via CATV networks (this may also be a satellite channel), where the PSTN is used as the return channel [16].

The heart of this system is formed by the Telecast® server, which manages the communication between the Internet and the users. The server, among other tasks, manages the current user addresses and provides user access to the Internet via a modem pool. Moreover, the server manages the user data and the establishment and monitoring of the connection via the PSTN. Finally, the server enables the CATV network operator to control the available transmission capacity. Depending on the specific requirements, transmission capacity can be guaranteed or can be allocated to different users dynamically. This is referred to as *dynamic subscriber management*.

The user, in turn, needs to install a Telecast® receiver with special hardware and software in his or her PC to receive the Internet services via the server. The receiver forms the interface between the PSTN, the CATV network, and the PC software. The application of dynamic subscriber management implies that the CATV network operator has full control over the receiver. Because the server can be accessed by users with a modem, as well as users with a Telecast® receiver, communication between both

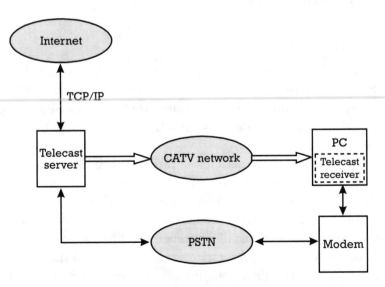

Figure 12.11 Fast Internet access via CATV networks. (*Source:* Telematic Systems & Services (TSS) B.V.)

types of users can be established. A bit rate of 7 or 8 Mbps can be achieved. If the DVB standard for digital transmission via CATV networks is used, the bit rate can even vary from 7 up to 41.3 Mbps (see Chapter 10). The combination of the telephone and the receiver can be a cost-efficient alternative to a modem.

One can think of all different kinds of fast Internet services that can be supported by this system. For example, real-time communication and software downloading are services that typically benefit from fast Internet access. This system can also be used for data broadcasting. Other examples are electronic publishing and multiplayer games.

12.6   Interactive services via teletext systems

Teletext systems, which are mainly used in Europe, can be integrated in an interactive system to provide different interactive services to the end user. Chapter 9 describes the teletext system. This section discusses the teletext system in a more service-oriented way, providing the reader with a convenient introduction before discussing the provision of interactive services via teletext systems.

12.6.1   Teletext systems

A television signal includes several lines that are together called the VBI. Beside data broadcasting, the VBI can, for example, be used for teletext. In Europe, the required teletext decoders are incorporated in television sets. Consumers can choose between a television set with or without a teletext decoder.

The teletext pages are placed in a carousel. Just like slides in a slide projector, these pages are broadcast in a fixed order via a teletext transmitter. The average time between the display of two different pages is 12.5 seconds. The user can select a maximum number of 800 main pages. Every main page can consist of a number of subpages. For example, the indication "3/4" may be presented in the upper part of the screen. This implies that this main page contains four subpages and that the user is watching the third subpage. The application of subpages allows a dramatic increase in the total number of pages. At the same time, however, it implies an increase in the time between the display of two different pages. Today, there are systems on the market that provide 32,000 pages.

By selecting a page, the user can access information free of charge. These pages can be constructed by editors who collect information themselves or by information service providers. In the latter case, the information is often transformed into teletext pages automatically. As a teletext page consists of 24 lines of 40 characters each, graphical objects can be displayed with limited quality only. Hence, the information services are often text-oriented. Examples of information services, which are typically provided via teletext systems, are television guides, news, weather forecasts, flight information, and stock market exchange information.

12.6.2 Interactive services

Teletext systems can be used to construct a system by which interactive services can be provided. Despite the limited graphical quality, messaging services like e-mail are possible. Moreover, retrieval services such as access to external databases can be supported. Figure 12.12 shows a functional description of interactive service provision via a teletext system.

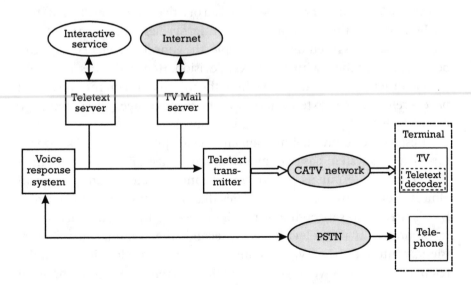

Figure 12.12 Interactive services via a teletext system. (*Source:* Telematic Systems & Services (TSS) B.V.)

The user controls a terminal that consists of a television set with a built-in teletext decoder and a telephone with push buttons. The user calls a voice response system via the PSTN. After word of welcome to the concerned interactive service, this system assigns a teletext page to the user. The voice response system is controlled by the user via his or her telephone's push buttons. Alternatively, a special keyboard can be used to send ASCII characters.

As soon as the connection with the interactive service provider is established, the user can request information. The information is provided to a teletext transmitter and displayed on the screen directly. This is an important difference from the normal application of a teletext system, where a carousel with teletext pages is broadcast to the user. The heart of the system is formed by a teletext server, which controls the information flow during the interactive session. The teletext server communicates with all parts of the system and manages the assignment of the free teletext pages. Examples of interactive services that can be provided via this system are personal insurance information, flight bookings, and stock trading.

Another type of interactive service that makes use of the teletext system is TV Mail®. By means of this service e-mails can be sent and received (off-line) via the Internet and be presented on the television screen. In this case, a TV Mail® server is used instead of a teletext server. This system enables one of the most important Internet services (e-mail) to be accessed via a mass medium. In principle, the concerned teletext page can be watched by others with a teletext television. Hence, there is a possibility that others can determine to whom the information on the concerned page is related. If the extension code of the teletext page is used, this page can be watched by the individual user only.

Despite the graphical limitation and the telephone's limited keyboard function, this system has two important advantages. First, it offers a high penetration of terminals (television sets and telephones are present in almost every household). This implies that, in contrast with the current DVB systems, no set-top box is required because the teletext decoder is incorporated in the television set. Second, the terminal has a user-friendly interface. (Everybody knows how to use a television set and a telephone.) These advantages may be the difference between long-term and short-term return on investment in interactive services that are provided via the television medium.

12.7 Summary and conclusions

DVB has produced several specifications for the provision of interactive services via broadcast networks. This requires interactive systems that at least make use of a return interaction channel from the user to the service provider. In addition, a forward interaction channel from the service provider to the user may be used. The DVB specifications concern the interaction channels for CATV networks and the interaction channels through PSTN and ISDN. These specifications are currently being standardized by ETSI.

The work of DVB is based on the use of a set-top box. However, there are alternative systems to provide interactive services via broadcast networks, which do not require a set-top box. For example, Internet services can be provided via broadcast networks to the user by means of a Telecast® server. In turn, the user needs to install special hardware and software in his or her PC, and a modem is required to establish the return channel via the PSTN. Another example is the provision of interactive services via a teletext system. This requires a teletext decoder, which is built into the television set, and a telephone with push buttons. Despite the graphical limitations of the teletext system, interesting text-oriented services such as personal insurance information and TV Mail® can be provided in a user-friendly way.

References

[1] DVB, *Commercial Requirements for Asymmetric Interactive Services Supporting Broadcast to the Home with Narrowband Return Channels*, DVB document A008, October 1995.

[2] EBU/CENELEC/ETSI-JTC, Digital Video Broadcasting (DVB); *Network Independent Protocols for Interactive Services*, prETS 300 802.

[3] DVB-SIS, *Guidelines for the use of the DVB specification: Network Independent Protocols for Interactive Services*, 5 July, 1996.

[4] EBU, "Functional Model of a Conditional Access System", *EBU Technical Review*, Winter 1995, pp. 13–14.

[5] Shanmugam, K. S., *Digital and Analog Communication Systems*, John Wiley & Sons, 1985.

[6]   EBU/CENELEC/ETSI-JTC, Digital Video Broadcasting (DVB); *DVB interaction channel for Cable TV distribution system (CATV)*; TM 1640 Rev.5, prETS 300 800, 30 September, 1996.

[7]   ITU, *ITU-R Recommendation I.361 for ATM UNI*.

[8]   Ha, T. T., *Digital Satellite Communications*, New York: Macmillan Publishing Company, 1988.

[9]   EBU/CENELEC/ETSI-JTC, *Digital Video Broadcasting (DVB); Interaction channel through PSTN/ISDN*, prETS 300 801, December, 1996.

[10] ETSI, *Attachments to the Public Switched Telephone Network (PSTN); General technical requirements for equipment connected to an analogue subscriber interface in the PSTN*, ETS 300 001 (NET 4).

[11] EBU/CENELEC/ETSI-JTC, *Interface for DVB-IRD*, prEN50202.

[12] Council Directive 91/263/EEC on the approximation of the laws of the Member States concerning telecommunications terminal equipment, including the mutual recognition of their conformity, O.J. L128, 23 May, 1991.

[13] ETSI, *Integrated Digital Services Network (ISDN): Basic rate user-network interface Layer 1 specification and test principles*, ETS 300 011.

[14] ETSI, *ISDN user-network interface/Data link layer specification*, 1994 and *ITU-T Recommendation Q.921 Rev1*.

[15] ETSI, *Digital subscriber Signaling System No.1 (DSS 1) - ISDN user-network interface layer 3 specification for basic call control*, 1994 and *ITU-T Recommendation Q.931 Rev1*.

[16] Bons, J. H., "Internet via de kabel op de PC en TV," *Telematica Nieuwsbrief*, No. 3, March, 1997.

Contents

European digital video broadcasting analyzed

13.1   Introduction

This chapter analyzes the European DVB framework by means of the *technology assessment* (TA) concept. This framework includes the DVB project results as described in Chapters 8–12 and the EU policy on DTV, as explained in Chapter 6.

As will become clear, the TA concept provides a useful instrument for the purpose of this analysis, in which the analytical model (see Chapter 7) plays an important role. The different actors' possible roles (i.e., the roles of government as well as market parties) on a mid-term time scale in the introduction of interactive DTV services, which are provided via a CA system, are discussed.

13.2   Technology assessment

This section explains the background, functions, and several different types of TA. Furthermore, the evolution of TA toward the establishment of an integral technology policy in general is discussed.

13.2.1   The technology assessment concept

In their study [1] on how TA can contribute to the advancement of decision-making on technological developments in politics and policy, Dutch researchers Smits and Leyten stated that technological developments should not be regarded as exogenous determining factors, but rather as the product of activities and relationships within society as a whole. TA should be seen as the response to an increasing necessity to socialize decision-making on technology. Smits and Leyten defined TA as follows:

> Technology assessment is a process consisting of analyses of technological developments and their consequences, plus the discussions in response to those analyses. The goal of TA is to provide information to those people involved with technological development in order to help them establish their strategic policy.

TA can be used as an instrument for different objectives. Concerning the TA concept's practical implementation, eight functions can be distinguished:

1. Strengthening the position of actors in the decision-making process;
2. Facilitating short- and mid-term policy;
3. Contributing to the development of long-term policy;
4. Early warning (especially for negative consequences);
5. Broadening the scope of knowledge and decision-making concerning actors;

6. Tracking and developing socially desirable and beneficial technological applications;

7. Stimulating the acceptance of technological innovation by the
   public;

8. Stimulating scientists' awareness of their social responsibility.

Three main types of TA, the result of developments of technology
policies from the past, currently exist in practice:

▶ Reactive, early warning TA;

▶ Active, oriented on supporting the current policy TA;

▶ Active, oriented on the initiation of new long-term policy TA.

The primary objective of reactive, early warning TA is to provide
information on the possible negative consequences of technological
developments. This type of TA emphasizes functions 4 and 8. The second
type's objective is to support the (technology) policy of parliaments,
(parts of) the government, and social groups (e.g., trade unions) by providing information. This enables the evaluation of other parties' proposals, to search within certain limits for alternatives, and to support their
own proposals. This also includes the development of concrete and specific technological applications. Functions 1, 2, and 6 are emphasized. The
last type of TA is used to develop possible scenarios for developments in
society in which technology plays a special role. By doing so, attempts are
made to discuss technology-related subjects that are relevant for policy
making. Functions 3 and 5 are stressed by this type of TA.

13.2.2 Technology assessment in an integral technology policy

Smits and Leyten make a plea for a further evolution of TA. They believe
that TA should be embedded in an integral technology policy that, by
means of research, the establishment of networks, advice, and the initiation of discussion, contributes to the following:

▶ The generation of knowledge on stimulating the awareness of
  social, economic, and material options that relate to technological

developments—this with the objective to facilitate the demand's articulation;

▶ Stimulating the debate on the direction of technological developments in relation to social-institutional questions, by which means a better balance between the characteristics and potentials of the society concerned and the technological developments can be achieved;

▶ The development by actors of a technological as well as a socio-institutional innovative strategy. This must serve as a basis for the development of ideas on significant and/or desired applications.

In other words, TA tries to contribute to optimally adapt the techno-economic system to the socio-institutional system. It is assumed that both systems interact on an equal basis. This implicates that the socio-institutional system does not necessarily lag behind the techno-economic system. TA can be considered an institutional change agent that constantly tries to fill the gap between both subsystems at a strategic level. By considering technology as a source of socio-institutional options, TA can even contribute to give social and political innovation a leading position in techno-economic developments. With respect to an integral technology policy, three forms of TA can now be considered:

▶ *Awareness TA* (ATA);

▶ *Strategic TA* (STA);

▶ *Constructive TA* (CTA).

The primary function of ATA is monitoring the potential of technological developments and creating awareness of the options for society that relate to these potentials. The other way around, monitoring the social developments and creating awareness of the technological options, is also considered as part of the ATA domain. The level of analysis is global and less specific and has a time scale of ten years minimum. It concerns studies on long-term developments and the related options.

STA aims to facilitate the strategy- and consensus-building process by specifying the global-level ATA results to a sector or an actor. The precondition for an effective STA system is in close relation with the concerned policy domains and sectors in society. An intensive interaction is required to evolve from the technological potentials to an application-oriented strategy.

CTA's goal is to strengthen the ties between the process of technology design and the area of application of new technologies. This concerns the establishment of conditions to allow significant learning and searching processes in the diffusion phase and effective feedback to conduct research and development and product development.

Each of these TA forms has its own specific function, allowing it to facilitate the policy-making process, depending on the requirements at different levels of decision making. As stated in Section 13.1, this chapter aims to assess the different actors' possible roles on a mid-term time scale in the introduction of interactive DTV services that are provided via a CA system. This assessment has the objective of facilitating the strategy- and consensus-building process by specifying the DVB results of a sector or an actor. Hence, this assessment can be characterized as a form of STA.

13.3 Results of the European digital video broadcasting

For the purpose of the above mentioned STA, this section discusses the results of the DVB project and the EU policy on DTV.

13.3.1 The DVB project

The DVB project has produced guidelines on source coding and multiplexing of television signals. Moreover, DVB has specified several television broadcasting systems for transmission via satellite, cable, and terrestrial networks. A highlight was the development of a European standard on CA, which occurred following a consensus model. Market parties took the initiative in producing the concerned specifications. Most of the participating market parties had direct commercial interests. The national governments and the European Commission participated from a different perspective: the structuring of the market in order to establish a level playing field. As a result, DVB specified the common scrambling algorithm (i.e., a common crypto system). DVB decided that the CAMS was not to be subject to standardization. Hence, the CI was specified. In addition, specifications on Transcontrol and Simulcrypt were developed. By specifying several return channels, DVB entered the telecommunications (policy) domain.

In dealing with the common scrambling algorithm, the concerned national government agencies constrained this crypto system's complexity to facilitate their ability to lawfully intercept communications to protect public order and national security, while retaining the ability to protect intellectual property rights. It should be noted that the European Commission has no competence in the field of public order and national security.

In addition, an administrative measure has been introduced in order to counter piracy and the abuse of the common scrambling algorithm by criminals, terrorists, and distrusted foreign governments. ETSI acts as custodian of the specifications on the crypto system's encryption part. This implies that ETSI issues these specifications, after the concerned market party has signed a licensing agreement, and registers the licensee. This agreement includes a nondisclosure agreement. As part of the procedure, ETSI checks whether an applicant has ever been involved in piracy activities in the past. If this is the case, it will not issue a license.

13.3.2 EU Directive on television standards

DVB has proven itself capable of providing digital technology for television viewers. This was also recognized by the government. Through the cooperation of national governments and the European Commission, the Television Standards Directive [2] was accomplished. On July 25, 1995, the EU's Council of Ministers, unanimously approved the Directive and established the Directive on October 24 of that same year. Consequently, the specifications developed in the DVB project and made into standards by ETSI were made obligatory. These specifications mainly concern the norms for generating program signals and the adaptation to the transmission media of satellite, cable, and terrestrial networks. This generates an important political embodiment for digital (wide-screen) television.

The European Commission also intends to structure the market for DTV services such as pay-TV by using this Directive. For example, for the encryption of television programs the use of the common scrambling algorithm is mandatory. Furthermore, digital (wide-screen) television sets must be provided with a CI, which, however, is not mandatory for the set-top box. Also, the Directive regulates network access to DTV services (Simulcrypt). This restricts the positions of, among others, BskyB and Canal Plus. Furthermore, CA systems that are exploited on the market must dispose of the necessary technical possibilities for an inexpensive conveyance of control to the cable head-ends (Transcontrol). Herewith the CATV operators must be able to have complete control on a local or regional level over the services that use such systems for CA. Accordingly, the CATV operators can enforce the actual use of Transcontrol on the grounds of this Directive.

There are also stipulations included in the Directive that aims to realize a level playing field on the grounds of the community competition rules. Herein the opposing of dominant positions on the market is clearly stated. In addition, licenses concerning DVB specifications must be issued on grounds of nondiscriminatory bases, so that no barriers arise for new entrants to the market. Subsequently, the Member States must provide arbitration procedures to settle unsolved disputes honorably, timely, and transparently.

13.3.3 (draft) EU Directive on the Legal Protection against Piracy

The key issue in providing information services via CA systems is to ensure that only authorized users (i.e., users with a valid contract) can get access to a particular programming package. Encryption is often used as a means of technically protecting these services from unauthorized access (piracy). In some Member States the legal protection against piracy is insufficient or even absent. This could lead to a wide-spread use of illicit devices—i.e., equipment or software designed or adapted to give access to a protected service in an intelligible form without the service provider's authorization. DVB recognized this problem from its beginning when installing the ad hoc group on CA (see Chapter 8). One of this ad hoc group's tasks was to make recommendations for the necessary flanking pan-European policy to discourage piracy. However, it was the intellectual property rights ad hoc group that produced the DVB recommendations on antipiracy legislation[3]for DVB in October 1995. The European Parliament shared DVB's concerns and amended the Television Standards Directive with a recital to establish an effective Community legal framework on antipiracy. This recital was adopted by the council when establishing this Directive on October 24, 1995.

In March 1996 the European Commission published the green paper "Legal Protection of Encrypted Services in the Internal Market" [4]. Its preceding wide-ranging consultation confirmed the need for a Community legal instrument ensuring the legal protection of all those services whose remuneration relies on CA. As a result, the European Parliament and the Council proposed a Directive on the "Legal Protection of Services Based on, or consisting of, CA" [5]. The current (May 1998) draft Directive provides an equivalent level of protection between member states relating to commercial activities that concern illicit devices. However, this draft Directive's implementation may not result in obstacles in the internal market concerning the free movement of services and goods.

The protected services concern radio, television and information society services (e.g., video-on-demand, games, and teleshopping) and the provision of CA to these services as a service in its own right. As such, this draft Directive prohibits and sanctions the manufacture, import, distribution, sale, rental, possession, installation, maintenance, or replacement for commercial purposes of illicit devices. Moreover, the use of commercial communications to promote illicit devices is prohibited and sanctioned. The draft Directive explicitly does not cover the private possession of illicit devices, intellectual property rights, the protection of minors and/or national policies on the protection of public order or national security. The latter implicates that lawful interception as part of a national policy on cryptography is not considered as piracy.

13.4 Analytical model and digital video broadcasting

After describing the results of the European DVB framework, the analytical model (see Chapter 7) will now serve as a useful tool for the purposes of the actual STA. Analogous to Chapter 7, the conceptual model is used to visualize the impact of these results. Preceding to the analysis, a vision on an integral technology policy on DTV is discussed.

13.4.1 An integral technology policy on digital television

The analytical model describes the various aspects involved in the development of DTV services that are provided via a CA system. A conceptual model is used to visualize the relationships between this technological development, its consequences, and the eventual desired situation. One could state that, by means of the analytical model, the technology's potentials are identified in order to create awareness about the options for the various actors in society. The leading principle of an integral technology policy in this field is that the above-mentioned services cannot be developed successfully (i.e., embedded in society) without recognizing and addressing the socio-institutional, as well as the techno-economic aspects.

13.4.2 The conceptual model

As stated in Chapter 7, the conceptual model describes the relationships between technical developments and the consequences of these developments. Moreover, this model identifies per consequence what aspects need to be taken into account in order to responsibly embed these technologies in society. For society, it is important that:

- Services with a great social and economic interest are available;

- The information's multiformity is assured in the case of a limited and one-sided supply of services;

- The affordability of basic services is guaranteed if high costs lead to haves and have nots;

- An efficient and effective cost allocation in the economic value-added chain is achieved in case costs are shifting to different layers in the economic value-added chain;

- An open market structure is established despite the cost increase due to the application of a CI;

- A one-stop shop is created in the situations where many CAMSs (i.e., modules) are used;

- Privacy is protected in case personal data is registered;

▶ If strong crypto systems are used, lawful interception is assured to protect national security and public order. On the other hand, the crypto system must be strong to protect intellectual property rights against piracy. Lawful interception implies that privacy protection is limited;

▶ Intellectual property rights are (legally) protected, in case a CA system is subject to piracy.

From the perspective of an integral technology policy on DTV it can be stated that the functions (see Section 13.2.1) 1 to 3, 7, and 8 have a socio-institutional character. The aspects 4 to 6 and 9, on the other hand, are to be characterized as more techno-economic. Moreover, this assessment leads to the conclusion that aspect 7 incorporates a conflict between the socio-institutional and techno-economic systems. All these aspects have to be addressed to responsibly embed these technologies in society.

The European DVB framework only covers part of all aspects. The fifth aspect has been addressed by the specification of the DVB CI. Despite the cost increase, the EU Television Standards Directive mandates that digital (wide-screen) television sets must be provided with a CI. Note that the CI is not mandatory for digital set-top boxes! Next, aspect 8 has been addressed in that the four concerned market parties, after consultation with several export control-related authorities, have specified the DVB crypto system for CA to be applied within digital broadcasting systems. In addition, the DVB crypto system (including the actual algorithm as well as the key length) is kept confidential. For this purpose companies (that have not engaged in audio-visual piracy) have to sign the *nondisclosure agreement*. According to the EU Television Standards Directive, the use of the CSA in consumer equipment for CA is mandatory. Moreover, the Directive requires that the CSA is administered by a recognized standardization body. The *DVB Descrambling Custodian Agreement* and the *DVB Scrambling Technology Custodian Agreement* specify these terms of administration with ETSI, acting as neutral custodian for the CSA specifications. Finally, function 9 has been addressed through the EU initiative on the legal protection of CA against piracy.

The conceptual model, presented in Figure 13.1, provides an overview of the functions (indicated in gray) that are covered by the European DVB framework.

CA: conditional access
CAMS: conditional access management system
CI: common interface
CSA: common scrambling algorithm
IDTV: interactive digital television

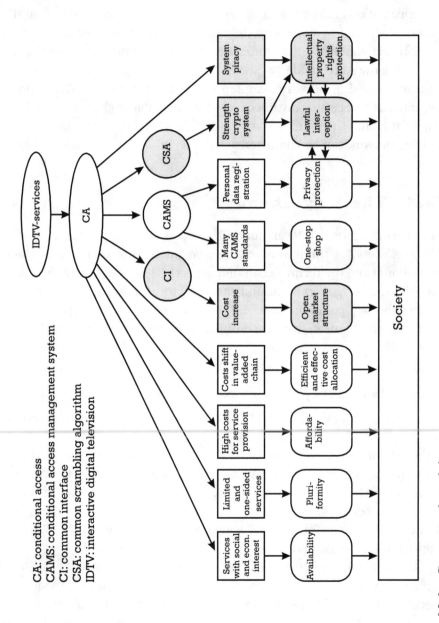

Figure 13.1 Conceptual model.

As can be concluded from the conceptual model, several important aspects have remained outside the European DVB framework's scope. With regard to the techno-economic functions, these concern an efficient and effective cost allocation in the economic value-added chain (4) and the establishment of a one-stop shop (6). Although the Television Standards Directive aims to structure the market by, among other things, mandating the concerned ETSI standards and defining competition rules, functions 4 and 6, however, are left to the market parties. The socio-institutional aspects that are not covered are availability (1), multiformity (2), affordability, (3) and privacy protection (7). Hence, the European DVB framework can be characterized as a techno-economic system.

13.5 The way ahead

As stated once more, DTV can not be developed successfully (i.e., embedded in society) without recognizing and addressing both the socio-institutional and the techno-economic aspects. This section describes the possible roles of government and market parties in achieving an integral technology policy in this field. For this purpose, the aspects that have remained outside the scope of the European DVB framework are discussed.

13.5.1 Availability

It is important that the availability of services with a great social and economic interest (e.g., electronic elections or referenda) is guaranteed. One of the instruments governments use to ensure the availability of services with such an interest is that, as part of a broadcasting license, a network provider must carry the specific service to all consumers who are connected to its network. This instrument, which is referred to as a *must carry*, is often used to ensure national television broadcasting via CATV networks and terrestrial networks.

Another well-known government instrument is that, as part of a telecommunications license, a network provider is obligated to provide its specific service to all consumers within a certain geographical area (i.e., country) based on the same conditions and on the same tariff. Moreover, every consumer within this geographical area has to be connected to the network on his or her request, even if this leads to extra costs. This instrument is referred to as *universal service provision*. Government licenses for national telephony services often include the obligation for telecommunications operators to provide these services based on universal service provision.

13.5.2 Multiformity

A great variety of information, from various sources and from different perspectives is of great interest to society. The information supply should meet the requirements from several groups of the population and should contribute to the development of all individuals in society. Hence, the multiformity of information is an important social requirement. For this purpose, three different government policy instruments could be used:

- *Self-regulation* through financing via the price mechanism and limiting the government's role to creating conditions;

- *Product financing*, in which financing partly takes place via the price mechanism and partly via financing specific products by and/or because of the government;

- *Supplier financing*, in which the government assigns one or several information providers that have to take care of a multiform information supply.

In practice, these instruments can be combined, where different weighing factors play a role in emphasizing a specific government policy. These factors concern the confidence in the market to establish a multiformity of information services by itself, the need for demanding guarantees for a multiform information supply, the extent of government interference (especially in the case of product financing), and the possible unintentional side-effects and practical implementation.

13.5.3 Affordability

A service's availability does not necessarily imply its accessibility. In addition to a user-friendly interface, this can also depend on the affordability of the service concerned. The government's application of the must carry and universal service provision instruments for the purpose of availability also leads to economies of scale (as a side-effect), which, in turn, could lead to a price decrease. A more specific instrument that could be used by governments, however, is the individual subsidization of citizens with less financial means. This instrument's advantage is that its impact on the market structure is less significant. Moreover, it can be used independently from other government policy instruments.

13.5.4 Cost allocation

The introduction of interactive DTV services, which are provided via CA systems, requires large investments. It should be avoided that, due to new options as a result of an open set-top box, costs are shifted to other layers in the layer model in such a way that thresholds arise for market parties to invest in these new technologies.

It would, for example, be cost-effective and efficient if the network service provider centrally controlled the SAS, rather than the information service providers controlling their own proprietary SAS. This would reduce costs for the established pay-TV broadcasters, which have not invested in digital pay-TV yet, as well as for other (new) information service providers. This model allows the current information service providers to (still) control their own SMS, while new market entries do not have to invest in such a system. It could even be possible to develop a standard for the SAS. In the end, this results in economies of scale, so that the user pays less.

This model requires that information service providers pay the network service provider for the value-added SAS service. In addition, they should use a separate bookkeeping for activities in the different layers of the layer model. This implies that the value-added SAS service should be separated from the network service in the books. This is because the network service provider may want to provide information services on the basis of CA as well, while using the same SAS. Moreover, depending on the situation, the information service provider may have to pay the concerned network service providers for the transport (i.e., broadcasting) of their information services via their networks.

13.5.5 One-stop shop

Several information service providers (i.e., pay-TV broadcasters) that have integrated vertically on the basis of cooperation with CATV network operators have already announced their willingness to provide a one-stop shop. Next to these efforts, various CATV network operators intend to manage a counter through which (their own) services can be provided.

The use of Simulcrypt and Transcontrol has been agreed upon within the DVB project. These agreements are adopted in the Television Standards Directive. In practice, Simulcrypt facilitates the common use of an established pay-TV broadcaster's DTV broadcasting network by other information service providers with their proprietary CAMSs. Hence, a one-stop shop managed by an established information service provider is feasible. The use of Transcontrol, in turn, allows the network service providers (i.e., the CATV network operators) to control the CATV network's management. This facilitates a one-stop shop managed by CATV network operators. However, no agreement has been reached on the conditions under which Transcontrol will be applied. One scenario calls for the SMS to subside under the management of the information service provider, while the SAS is managed by a different party—for example, the CATV network operator. It should be noted that Simulcrypt and Transcontrol do not exclude each other. A CATV network operator can always decide to apply Transcontrol, even if the information services are provided via Simulcrypt. Hence, Transcontrol may be used in addition to Simulcrypt.

There will probably be some information service providers, as well as network service providers, that will manage a one-stop shop. As a result, users within a specific geographical area can choose between a limited number of shops that provide several services. This can be compared with supermarkets that supply all different kinds of products. From the user's perspective, every shop should only issue one smart card that enables the user to access the information services from different information service providers through that shop. This avoids the situation in which a new smart card is required for each service that is provided.

Ideally, the one-stop shop's management is cost-efficient and is independent of the information service providers. In this way, it might garner a level playing field for the current players as well as new market entrants. Nondiscriminatory access to networks is an essential condition to ensure a level playing field. Finally, it is undesirable that a specific type of information service (e.g., the downloading of dedicated software) is exclusively included in the package of a single one-stop shop. This could, again, lead to the protection of a dominant position in the market. Governments should apply their competition instruments, whenever appropriate, to guarantee an open market structure.

13.5.6 Privacy

If personal data (i.e., individual consumptive behavior) is registered by service providers for purposes other than billing or statistical analyses, the consumer's privacy may be affected. This especially holds true for the provision of interactive services.

In Europe privacy rules are laid down in several laws and regulations. Article 8 of the European Treaty on the Protection of Human Rights and the Fundamental Freedoms [6] states that every human being has the right that his or her private life, family life, house, and correspondence are to be respected. The general principles on privacy protection against automatic processing of personal data are laid down by the Council of Europe in the Strasbourg Treaty [7]. Moreover, the Council of Europe made a specific telecommunications privacy recommendation [8], and the European Commission established the EU Privacy Directive [9] and the EU Telecommunications Privacy Directive [10]. As DVB entered the telecommunications (policy) domain by providing interactive services (i.e., return channels), the latter is also applicable to interactive DTV services. This Directive mandates, among other things, that client data is only to be collected for billing purposes and is to be passed to third parties only if the data is anonymous.

The EU Privacy Directive provides for the establishment of privacy codes. A privacy code can be regarded as a sector-specific self-regulatory instrument that can be applied within the boundaries of the existing privacy legislation. The objective is to make agreements within a sector on the handling of personal data, while protecting consumers' privacy. As the sector itself formulates the privacy code, the advantage is that it meets the concerned sector's requirements better than if governments were to define these rules. A privacy code is considered a quality feature by the sector and, as such, can also be used for marketing purposes.

The market parties involved in interactive DTV, especially information service providers, could consider developing a privacy code in this field. Ideally, this code would be developed by DVB itself, rather than by various groups independently. At the same time, this would allow market parties that provide their services outside the EU to comply to the same privacy code and, hence, refer to the same quality standard. The privacy code could very well be included in the memorandum of understanding. Alternatively, the DVB members could sign a privacy code separate from the memorandum of understanding on a voluntary basis. However, the impact of the latter would be less significant. The following list presents several guidelines that play an important role in such a privacy code.

1. Definitions: Personal data is data that can be reduced to an individual natural person. A personal data registration is a coherent collection of personal data, relating to several individuals' personal data, that is managed via automatic means or, with regard to an effective retrieval of that data, has been established systematically. A keeper is someone who controls the personal data registration.

2. Personal data should be collected and processed lawfully.

3. Registration of personal data should only take place for the purpose of billing, unless the user explicitly authorizes the keeper to use this data for other purposes (opting-in-principle).

4. The interrogative sentence in the case of opting-in should occur with every service and should be stated clearly and unequivocally.

5. Other purposes than billing should be stated with the user's application (purpose-binding-principle).

6. The keepers of personal data registrations should notify the registries by correspondence within one month about the registration of their personal data in a personal data registration.

7. Personal data should only be handed to those parties that have been indicated with the user's application.

8. Only data that can not be reduced to individuals (e.g., statistical data) may be handed to third parties without prior authorization by the users.

9. Personal data should be accurate and, if appropriate, should be kept up-to-date.

10. Personal data should only be relevant for the purpose of its collection.

11. Registries should be able to retrieve and look at their dossiers and have them corrected or deleted.

12. Personal data should be kept no longer than is necessary for the purpose of its collection.

13. Only the keeper has access to the personal data and has the duty to keep the data secret.

14. Appropriate security measures should be taken in order to secure the personal data registration.

15. Keepers of personal data registration should register themselves with the Data Protection Registrar.

16. Registries should be able to file complaints with the concerned service provider and an appropriate body.

17. A fast and thorough investigation by the Data Protection Registrar should be possible.

In addition to a privacy code, the implementation of Transcontrol, which allows the information service provider to manage its own client data and in which the network service provider controls the transport of DTV services, deserves further elaboration in the context of privacy. In the driving of the SAS by the SMS, for instance, it could be considered to make the client data anonymous. It is very much possible, technically, to instruct the SAS without an assignment that can be traced back to an individual person. This is beneficial to both the service providers (for client-data protection) as well as to the consumers (for privacy protection). Hence, this data must be handled confidentially!

13.6 Summary and conclusions

The TA-concept proved to be a useful instrument for analyzing the European DVB framework. This framework includes the DVB project itself and the EU policy on DTV—in particular the Television Standards Directive and the (draft) Directive on the Legal Protection against Piracy. The leading principle in this analysis has been an integral technology policy on interactive DTV services, which are provided via a CA system, in which the socio-institutional aspects, as well as the techno-economic aspects are addressed in order to responsibly (i.e., successfully) embed this technology in society.

It can be concluded that the market-driven DVB project has proven to be very successful with respect to the aspects that fall within this project's scope. However, the DVB project, as well as the concerned EU policy, mainly focused on aspects with a more techno-economic character and did not address the conceptual model's socio-institutional factors other than lawful interception. Hence, the European DVB framework can be characterized as a techno-economic system.

With regard to the factors that are not addressed by the European DVB framework, the possible roles of governments and market parties on a mid-term time scale in the introduction of this technology have been discussed. Concerning the techno-economic aspects (an efficient and effective cost allocation in the economic value-added chain and the establishment of a one-stop shop), which were not covered, the market parties' role is emphasized. In the case of the not covered socio-institutional aspects (availability, multiformity, and affordability) the government's responsibility is stressed. An exception is made in the case of the socio-institutional privacy protection factor. Although several (telecommunications) privacy laws and regulations apply at a more general level, the DVB project's members could prove its strength by establishing a privacy code on interactive DTV services themselves. This code's objective is to ensure privacy protection in further (technical) implementations of this technology. Such a code should be considered a quality feature rather than a threat.

References

[1] Smits, R., and J. Leyten, *Technology Assessment: Watchdog or Tracker? Towards a Comprehensive Technology Policy*, TNO Studiecentrum voor Technologie en Beleid, Kerckebosch b.v., Zeist, 1991.

[2] Directive 95/47/EEC of the European Parliament and of the Council of 24 October 1995 on the use of standards for the transmission of television signals, O.J. L281/51, 23 November, 1995.

[3] DVB, *Recommendations on Antipiracy Legislation for Digital Video Broadcasting*, A006 rev 1, October, 1995.

[4] European Commission, *Green paper on the Legal Protection of Encrypted Services in the Internal Market*, Brussels, COM (96) 76, final, 6 March, 1996.

[5] Proposal for a European Parliament and Council Directive on the Legal Protection of Services Based on, or consisting of, Conditional Access, 23 February, 1998.

[6] European Treaty on the Protection of Human Rights and the Fundamental Freedoms, Article 8.

[7] Strasbourg Treaty, Treaty of 1991 of the Council of Europe on the protection of individuals with regard to the automatic processing of personal data, 28 January 1981, Trb. 1988, 7, 28 January, 1981.

[8] Recommendation of the Council of Europe on the protection of personal data in the field of telecommunications services, Recommendation No.R(95)4, 7 February, 1995.

[9] Directive 95/46/EEC of the European Parliament and of the Council of 24 October, 1995 on the protection of individuals with regard to the processing of personal data and the free movement of such data.

[10] Directive 97/66/EEC of the European Parliament and of the Council of 15 December 1997 concerning the processing of personal data and the protection of privacy in the telecommunications sector, in particular the Integrated Services Digital Network (ISDN), and in the digital mobile networks, O.J. L 24, 30 January, 1998.

Contents

Future developments

14.1 Introduction

The development of television officially started in 1884, when the German Paul Gottlieb Nipkow patented his mechanical television system. Since then, electronic, color, high-definition, and digital television systems have been developed in various parts of the world. Several governments, including those of the European member states and the European Commission, the United States, and Japan, have also been developing policies and regulations in this field. As described in Chapters 4–6, each of these countries' policies have their own objectives and characteristics.

The previous chapters focused mainly on the European DVB framework. In addition to the technological and standardization aspects of the DVB project, the EU policy is analyzed as part of this framework. This chapter's objective is to describe some aspects of (possible) future developments in DTV. With reference to Chapter 13's analysis, two future scenarios for society, based on possible roles of governments and market parties, are described. Additionally, the fact that the migration toward the establishment of DTV and CA is not a straightforward process is discussed. Finally, the convergence process, which was identified in Chapter 3, is discussed from a technological and market perspective and from a policy and regulatory future perspective.

14.2   Two future scenarios for society

This section describes a pessimistic and an optimistic scenario on how interactive DTV services, which are provided via a CA system, are established on a mid-term time scale [1]. This bipolar approach allows us to clarify the boundaries within which the technological developments could take place. In reference to Chapter 13, only the aspects that have remained outside the European DVB framework's scope are discussed.

14.2.1   The pessimistic scenario

In the case of the pessimistic scenario it is assumed that neither the government nor the market parties will take any responsibility for the possible negative consequences of the technological developments.

14.2.1.1   Availability

It only seems feasible to create a return on investment in interactive DTV services, which are provided via a CA system, in geographical areas with a dense population. However, some of the provided services are of high economic and social interest. Because of the financial risk, investments are not made in other geographical areas. Concerning the services with a high economic and social interest per geographical area, different conditions and tariffs apply. The government does not create any incentives to stimulate the required investments. Moreover, the government is not willing to ensure equal conditions and the same tariffs for these important services, as this has major financial implications for the concerned market parties. As a result, the availability of services with a high economic and social interest remains low.

14.2.1.2 Multiformity

Only a few services providers are operating on the market. These service providers try to ensure as much return on investment as possible by mainly providing mass-entertainment programs. Hence, there is hardly a variety of information from various sources and from different perspectives. The government does not apply any financial instruments to create a multiform supply. In the end, only one-sided and limited types of services are being provided.

14.2.1.3 Affordability

Due to limited economies of scale, a return on investment can only be created by high prices that consumers have to pay for their services. The government is convinced that the market will simply do its work. Moreover, because there is no must-carry or universal service provision for services with a high economic and social interest, no economies of scale (as a side-effect) are achieved. Furthermore, individual subsidies to ensure that certain groups in society with less financial means have access to services that are necessary for their social functioning are not part of the government's policy. Hence, society is divided between haves and have nots.

14.2.1.4 Cost allocation

The established information service providers (i.e., pay-TV broadcasters) hold a strong position against network service providers and would to retain this position. However, the Television Standards Directive strengthens the CATV network operators' position through the possibility to enforce the use of Transcontrol. Moreover, this Directive mandates that the established information service providers give other parties access to their networks for digital subscriber services—for example, through the application of Simulcrypt.

A battle of power on the network management and the set-top box population's management takes place. This concerns, among other things, the multiplexing, program encryption, and control over the SMS. The assignment of responsibilities and the related cost shifts result in a concentration of costs in the layer model's network services layer. Furthermore, the government does not give any clarity to the market by establishing the required competition rules, such as a separate bookkeeping for the activities in the different layers within the layer model. As a result, investments are lacking, which implies that the current pay-TV broadcasters' dominant positions remain in place.

14.2.1.5 One-stop shop

The information service providers, as well as the network providers, want to create a counter through which their (own) services can be provided. There is a number of vertically integrated joint ventures between these parties, of which the information service provider manages the one-stop shop. Despite the principle of one-stop shopping, each service requires a different smart card. Moreover, other parties' access to the network is delayed or even refused, which prohibits the establishment of a level playing field. The government does not enforce nondiscriminatory access to the networks through the use of its competition rules. Because the government fails to create clarity in this matter, new dominant positions, with regard to the joint ventures mentioned above, are established.

14.2.1.6 Privacy

Individual consumptive behavior is registered for billing purposes and to create individual user profiles. Moreover, *sucker lists* that name individual consumers who have decided to buy a certain product right after its appearance in a commercial are created. Because of strategic concerns, this data is first being used internally for analyzing purposes. Next, the user profiles and/or sucker lists are sold to direct marketing organizations. Consequently, these organizations apply to these consumers to sell their products and/or services. The concerned organizations legitimize these marketing activities by stating that they can provide tailor-made products and/or services. The government does not enforce its privacy rules as would be required to protect the users' privacy. In the end, the users believe that their privacy is affected and loose their confidence in the concerned service providers as well as the government. Hence, the expected market growth is not achieved.

14.2.2 The optimistic scenario

In comparison with the pessimistic scenario, this scenario assumes that the government as well as market parties fulfill their tasks to responsibly embed DTV technology in society.

14.2.2.1 Availability

It is not commercially lucrative in all geographical areas to provide inter-active DTV services—some of which are of great economic and social interest—via CA systems. However, the drive for a bigger market share motivates market parties to provide services in geographical areas with a low population density as well. All together, a return on investment is achieved. Moreover, the government ensures that services with a great economic and social interest are provided under the same conditions and for the same tariffs.

14.2.2.2 Multiformity

Service providers of interactive DTV services, which are provided via a CA system, try to create a return on investment by providing for a services bouquet. The services' content is aimed at information and entertain-ment, as well as services that require explicit financial transactions (e.g., teleshopping and home banking). In addition, the government (partly) finances those services, which meet the needs of the different groups in society and contribute to the development of all individuals in society. By acting so, the government safeguards a multiform service provision.

14.2.2.3 Affordability

The market develops from pay-TV, via pay-per-view, toward video-on-demand and other interactive services. The pay-TV services are financed through (monthly) subscription fees. With pay-per-view, this comprises about 50% subscription and 50% billing charges for individual services. In the case of video-on-demand, the latter percentage will be higher. Because of this evolution, people are increasingly paying for consumed services only, rather than paying for the whole programming package as in the case of pay-TV. This creates transparency for the service providers, as well as the consumers, about the supply, the price, and the concerned service's success.

If the costs for services, which are of great economic and social inter-est, are too high, the government subsidizes individual citizens with less financial means. Moreover, the government ensures that these types of services are provided on the same conditions and for the same tariffs. A side effect is economies of scale that lead to a price decrease.

14.2.2.4   Cost allocation

The established pay-TV broadcasters have achieved a strong position in the market compared to the network service providers, because they were prepared to make large investments in the past. Currently, there are new market entries that are trying to develop their business in this field as well. Through the Television Standards Directive, the CATV network operators' position is strengthened, because the use of Transcontrol can be enforced. Moreover, the established pay-TV broadcasters have to give access to their DTV broadcasting networks—for example, by the application of Simulcrypt.

There is a common awareness of the mutual dependence, and, moreover, other parties are willing to invest as well. Hence, a balance between the various parties is achieved. This results in agreements on mutual responsibilities, especially concerning multiplexing, network management (notably SMS and SAS), the management of the set-top box population, as well as the related cost shifts. By these agreements, certainties are created, which means that the financial risks have decreased and that innovations take place. The government, however, requires a separate bookkeeping for cooperating market parties concerning the activities in the different layers within the layer model. Furthermore, the government ensures the non-discriminatory access to networks by means of its competition rules. The result is a level playing field.

14.2.2.5   One-stop shop

Information service providers, as well as network service providers, want to create a counter through which their (own) services can be provided. There are several joint ventures between both parties, where the CATV network operator manages the one-stop shop via the use of Transcontrol. On the other hand, satellite pay-TV broadcasters apply Simulcrypt to create a one-stop shop. In both cases (Transcontrol and Simulcrypt), only one smart card per one-stop shop is required to obtain access to the services in a geographical area. Other information service providers can supply their services through one or more of these shops, by which a level playing field is achieved. Furthermore, the government ensures the non-discriminatory access to the networks by a transparent use of its competition rules. Accordingly, dominant market parties do not get the chance to obstruct fair competition on the market.

14.2.2.6 Privacy

On the grounds of a mutual interest, the various service providers agree on a privacy code, because they are convinced that customer relationships are significantly dependent on the customers' trust in an adequate protection of their privacy. As such, this code is promoted as a quality feature in marketing strategies. Consequently, personal data is only used for billing purposes and anonymous statistical analyses. These statistical data may eventually be sold to third parties. The government's awareness of its primary task in privacy protection ensures the enforcement of privacy rules if necessary. The users' trust in the service providers, as well as the government, increases. This allows a return on investments to be achieved, and further innovations are stimulated.

14.3 Toward the digital area

The DVB project has provided several DTV systems that enable digital broadcasting via either satellite, CATV, or terrestrial networks. One of the important elements in DTV is the use of CA systems (see Chapter 11). These systems are used to ensure that only authorized users can watch a particular programming package. DVB developed three CA models: Multicrypt, Simulcrypt, and Transcontrol. This section describes these models and the ways in which they are adopted by the Television Standards Directive and discusses the migration toward the digital era.

14.3.1 The European DVB conditional access package

As concluded in Chapter 11, due to the various parties' different interests, CA has been an area of intensive debate in DVB. A balance had to be struck between opening up protected markets and at the same time not undermining the investments of the current, often vertically integrated, service providers. This has, among other things, resulted in the standardization of a common framework for the encryption of television programs, namely the CSA.

The DVB decided not to standardize the CAMS. An important reason was that the current service providers had already made big investments in their proprietary CAMSs, particularly in their SMS including the client data. Hence, a standardized interface between the proprietary CAMSs and the programs encrypted with the CSA was needed. For this purpose DVB specified the CI and, by this, supported the Multicrypt model. In this respect, the CI can be regarded as a solution to open up the market for horizontally oriented service providers—i.e., for competitors on a level playing field.

DVB has also produced a code of conduct for the nondiscriminatory use of Simulcrypt as well as the required technical specifications. This provides a model for the sharing of a transponder channel, so that in different geographical areas the same program can be received simultaneously by the set-top box populations of different service providers. This implies that vertically integrated service providers can no longer protect their market by excluding others from making use of their proprietary CA systems.

Moreover, DVB decided to support the use of Transcontrol. This allows the CATV network operators to control the services that use digital CA systems at a local or regional level. The use of Transcontrol is limited to CATV network operators only, rather than being applicable to all network service providers that directly provide television programs to consumers via their own CA systems.

The European Commission, in cooperation with several member states, has supported the work of DVB by establishing the Directive on Television Standards. This Directive, among other things, sets the mandatory standards for the CSA. The CI is mandatory for television sets with a built-in digital decoder. Note that the use of CI in set-top boxes is not mandatory! Hence, Multicrypt is supported in a "limited" way. Moreover, the Directive supports the Simulcrypt model by complying with the code of conduct. Finally, the Transcontrol model is adopted in the Directive but is limited to the case of CATV network operators. As DTV programs can be provided via all different kinds of networks to the end user (e.g., a PSTN or a terrestrial network), it would be feasible to apply Transcontrol as a more general principle. However, this has not been done.

14.3.2 Migration paths for digital television and conditional access

As explained in Section 14.3.1, there are three different models that can be used to establish a competitive market for CA. These models may even be combined. For example, a CATV network operator could always decide to use Transcontrol, even if the concerned programs are provided via Simulcrypt. That the migration from traditional analog broadcasting toward digital (pay-)TV is not a straightforward process is illustrated by the first digital satellite television broadcasting of free-TV channels in the Netherlands.

On July 1, 1996 the commercial broadcasting networks RTL4, RTL5, Veronica, and SBS6 started their first digital free-TV broadcasts in the Netherlands via satellite transmission. This required a set-top box for the conversion of digital signals to analog signals, so that these digital broadcasts could still be received with an analog television set. Because the consumer does not have to pay for these networks' programs, the set-top box does not facilitate a CA system. Hence, this a priori implies that neither Multicrypt, Simulcrypt, nor Transcontrol were supported. Moreover, because these networks then stopped their analog broadcasts via the Astra satellite on August 18 of the same year (the beginning of September for SBS6), this meant that the owners of satellite receivers were forced to purchase a new (digital) set-top box. The captains' protests in the Netherlands—they often receive television programs via satellite because of their ship's mobility—could not stop the digital era. The summary proceedings, started by the Consumers' Organization to force the networks involved to keep analog broadcasting beside digital (i.e., simulcasting), were settled to their disadvantage [2]. Apart from that, the CATV network operators, with their 95% penetration to households, were compensated by these commercial networks for the costs of installing an D/A conversion facility in the cable head-end. Hence, the CATV network operators' subscribers (i.e., most television subscribers in the Netherlands) can still watch television with their analog television sets.

The digital era cannot be stopped. However, there are still some impediments. When the terrestrial DTV broadcasts are introduced in the future, the aforementioned captains may, for example, wish to receive the programs via terrestrial digital networks. Is the newly purchased set-top box suitable for this purpose? Because these digital networks had not been introduced at the time, the required receiving facility was simply not installed in the set-top box. Moreover, this would have increased the set-top boxes' price, which is prohibitive to a lot of consumers anyway. If the captains would want to receive pay-TV in a later stage, they must buy yet another set-top box.

In order to achieve a high market penetration it is eminently important to keep the set-top boxes' price as low as possible. This holds true for commercial stations as well as for providers of pay-TV. However, making set-top boxes suitable for receiving via another transmission medium results in higher prices. To heighten the degree of penetration and at the same time discourage the protection of markets by only broadcasting via a particular transmission medium, every set-top box would have to be suitable for receiving signals via satellite networks as well as CATV and terrestrial networks. The question is whether or not consumers are willing to pay more for this freedom of choice. On the other hand, manufacturers could also exploit economies of scale to incorporate this facility as inexpensively as possible. These economies of scale can be achieved by, among other things, the application of the several digital transmission standards that were specified within the DVB project. Moreover, these standards have even been made mandatory by the Directive. Hence, the market could possibly regulate itself in this area.

As stated above with regard to CA, three different models on the provision of services based on CA can be used. Simulcrypt can be considered as the most worked-out model. DVB provided for technical specifications and the code of conduct is adopted in the Television Standards Directive. However, there are hardly any practical implementations of Simulcrypt on the market today. Next, Transcontrol is regarded as an important enabler for the introduction of one set-top box, if it takes place under conditions of nonexclusivity and nondiscrimination. However, no agreement has yet been reached on the conditions under which Transcontrol will be applied. The implementation of Transcontrol—which allows the service provider to manage its own client data (i.e., the SMS) and in which the provider of network services controls the clients' authorization (i.e., the SAS) and the transport of DTV services—deserves further elaboration. Additionally, in the driving of the SAS by the SMS it could be considered to make the personal data anonymous for privacy reasons. It is very much possible, technically, to instruct the SAS without an assignment that can be traced back to an individual person. For the service providers (client data protection) as well as for the consumer (privacy protection) the protection of this data is important.

The establishment of a healthy pay-TV market through the application of the Multicrypt model is addressed in a "limited" way. Because the CI is only mandatory for television sets with a built-in digital decoder and not for a digital set-top box, an additional development is required: the market introduction of digital (wide-screen) television sets for consumers [3]. Technically, this means that the digital decoder present in the set-top box is situated in the (wide-screen) television set instead. This gives the (now digital!) television set a connection (made mandatory by the directive) according to the CI's specifications. The set-top box will, in turn, also be provided with such a (standard) connection that it can be connected to the DTV set. Thus, a situation has been established where all the available services can be received with one set-top box.

The aforementioned captains, however, still do not have a digital (wide-screen) television set; they do have a digital set-top box with which they cannot receive pay-TV or any other service. According to the Directive a digital set-top box does not have to meet the CI's specifications, since in this Directive's philosophy the market will take care of the set-top box meeting the CI specifications at the introduction of digital (wide-screen) television sets. Because the average television set is replaced approximately every 10–12 years, migration via the Multicrypt model implies there is now a vacuum pending in the interim phase to a digital era. Therefore, this interim phase should be as short as possible. An improved (wide-)screen quality alone, however, is not enough to convince a lot of consumers to purchase a new digital (wide-screen) television set more quickly. The availability of alluring services could indeed accomplish this. New and current service providers will have to invest. The alternative is a decade in which the consumers lose faith in the service providers because they have to keep purchasing newly required set-top boxes. In order to prevent a "chicken or egg" situation, the development of new services will have to be combined with the introduction of digital (wide-screen) television sets. Because of their mutual dependence, but also in the interest of the consumers, manufacturers and service providers must take on this responsibility together as soon as possible. In the situation where the Multicrypt model is striven for, this is a negotiable, but longer, road to take from a self-regulatory perspective.

One can argue about what migration path is best to be used for the introduction of DTV. This also depends on the perception whether set-top boxes form a barrier to market penetration or not. The current reception of satellite television requires a set-top box for receiving and decoding purposes. Hence, consumers with a satellite set-top box probably will have a lower psychological barrier to (again) install such a device.

However, as described above, the introduction of set-top boxes without an open CA system can eventually lead to the loss of the consumers' confidence. Subscribers of CATV and terrestrial networks, on the other hand, can connect their analog television set directly to the network termination point. These consumers never needed a set-top box to receive their programs, unless they subscribed to a pay-TV broadcaster. In this respect, the advantage of a built-in digital decoder is that no set-top box is required for digital free-TV, as the conversion of DTV signals to the analog domain is now processed inside the television set itself. In this case, the Multicrypt model with its CI seems to be a logical step toward the establishment of digital pay-TV, but as the introduction of DTV sets is not foreseen in the near future yet, the (earlier) application of Transcontrol seams more feasible.

Another important aspect is the use of a feasible time scale to allow the market to synchronize the introduction of its products and services for DTV. Depending on the geographical area (the United States, the EU, or Japan), bodies like the FCC, the European Commission, and MPT, respectively, could consider publicly announcing a date on which analog television broadcasting will end. Ideally, these dates are set close to each other to foster an international level playing field. Considering the life cycle of current television sets and other television related equipment (e.g., video recorders) and to smooth the migration process, a simulcasting period is required. During the time between this announcement and the starting date for simulcasting, the various market parties can prepare their market introduction and the (national) governments have the time to allocate the required frequencies for satellite and terrestrial DTV services. If, for example, in 1999 it would be announced to the public that this intermediate period were to be three years and that simulcasting were to last 5 years, the digital era would start in 2002, and analog broadcasts would stop in 2007.

14.4 Convergence

The convergence process not only cuts through different technologies and traditional sectors but it affects the policy and regulatory domains of telecommunications and television broadcasting as well. This section illustrates the technological convergence perspective with a comparison between the Internet activities and the DVB project and discusses how the policy and regulatory frameworks may change as a result of technological convergence in this field.

14.4.1 Technology

As Chapter 3 describes, technological developments in information and communication technologies have led to a convergence of speech/audio, data, text, graphics, and video and thus to multimedia applications. This technological convergence also causes several traditional sectors to be subject to a convergence process. Chapter 3 describes this process for the traditional entertainment, information, telecommunications, and transaction sectors.

When looking at the DVB project itself in the light of a convergence with the developments concerning the Internet, at first sight both technological trajectories do not seem to have a lot in common. The current Internet has several characteristics that are clearly distinct from the DVB DTV services. For example, the Internet services are presented via a PC, rather than via the television medium as is the case with DVB. Due to the different presentation media, the types of networks that are used to transmit the concerning signals to the end user are also distinct. For DVB-developed specifications for digital satellite, CATV, and terrestrial networks, while Internet services in the beginning were provided via data networks and later also via telephony networks and other telecommunications networks (e.g., ATM and ISDN). Hence, the Internet can be regarded as a VAN in the telecommunications domain.

Next, the Internet can be characterized as a network with applications for and by users. Everyone who wants to share information with the world community places it on the Internet and, as such, determines (part of) an Internet service's content. In contrast, television's content is determined by content providers and information service providers.

Another important difference is the way standards are developed. Before going into detail on the Internet standardization process, it is necessary to understand some organizational background. In January 1992, the *Internet Society* (ISOC) was established. The ISOC is an international organization that aims to promote the use of the Internet for research and development and educational purposes. It provides a forum for governments, industry, and individuals to discuss regulation and procedures on the Internet. Moreover, the ISOC aims at the further development of

Internet technology and stimulates, among other things, the harmonization and standardization process within the Internet. In 1989, the current
standardization procedure became operational, characterizing itself with
a pragmatic approach to all parties concerned and its openness. This
openness is expressed through the fact that everyone can propose a new
standard, all documents are publicly available, and it is allowed to
develop new products based on these documents. All publications are
freely available online. The *Internet Architecture Board* (IAB) functions as
coordinating body. The *Internet Engineering Task Force* (IETF) is competent
in the technical areas. The IETF has several working groups that are
occupied with various technical subareas. Due to the large and still
fast-growing number of participants, the *Internet Engineering Steering
Group* (IESG) was established. The IESG consists of the working groups'
representatives.

Everyone is allowed to cooperate in the development of standards.
This takes place by the writing of a proposal in the form of a draft document. Other working groups can comment during a certain period.
Before a draft can be recognized as a standard, two working implementations based on the concerned draft have to be developed. The main difference with the DVB project is its pragmatic approach. A solution is often
developed based on a practical problem and put into practice in a testing
environment as soon as possible. This enables fast feedback, which allows
the standard to prove itself in practice already. This implies that the users
decide which products will be used. Hence, it is not so much Europe versus the United States and Japan, but rather political versus technical
solutions.

Finally, the Internet (still) has different users. The Internet is mainly
used by more highly educated people, among which are researchers and
scientists from research institutes and universities, students, and people
from knowledge-intensive industries and organizations. The DVB project
aims at advanced television for a broad audience, which is now expected
to behave as consumers. This is the information and entertainment sectors' traditional target group.

From the differences described in the paragraphs above, it can be concluded that the Internet activities and the DVB project are two trajectories with totally different characteristics. However, both trajectories are
increasingly becoming subject to a convergence process. As described in
Chapter 8, the specifications on interfaces to PDH, SDH, and ATM

networks are currently in the final stage of approval in ETSI. Moreover, an MHP is being developed. The MHP forms the API to all different kinds of multimedia applications. Finally, DVB is working on the transmission of data in DVB bit streams. This allows operators to, for example, download software over satellite, cable, or terrestrial links; to deliver Internet services over broadcast channels (using IP tunneling); or to provide interactive TV. Hence, the DVB (future) platform embodies the technological convergence between (traditional) telecommunications and broadcasting network and value-added services. This platform also overtakes the traditional PC versus television discussion. As the network and VAN services have converged, a distinction will now be made between different types of content. Hence, from the user's perspective one could think of a work-related display on the one end and a relaxation-related display on the other hand, both supported by the same technologies. From a technological point of view, the question that remains at this stage is whether this platform will mainly provide broadcasting services plus some telecommunications services or provide telecommunications services plus (still) some traditional broadcasting.

14.4.2 Policy and regulation

By providing a multimedia platform for interactive DTV services, rather than DTV broadcasting services only, DVB, in fact, entered the telecommunications (policy and regulatory) domain. However, as of now it still remains to be seen whether the (European) policy and regulatory frameworks on telecommunications and broadcasting will converge as well. If so, will both frameworks converge in such a way that the telecommunications framework will, in time, also fully incorporate DTV services—or perhaps even the other way around? It also remains to be seen whether these developments will lead to a completely liberalized television services environment as is the case with telecommunications. It is already possible to discern that public service television (of course its content is meant) will remain an important cornerstone of public (cultural) policy. Separate public funding structures will either remain in place or will be recreated to enable public policy influence in this area. In Europe, it is expected that the green paper on convergence will play an important role in the EU policy and regulatory environment to be developed for shaping the ICT revolution and, hence, the information society.

In this respect, it is feasible to establish two frameworks in the end. With regard to the layer model's functional distinction between content and transport (see Chapter 2), one framework could concern the information production, while the other could be an information transport framework. The latter typically includes telecommunications as well as broadcasting. This implies that there will no longer be a distinction between telecommunications and broadcasting policies and regulations and that the broadcasting sector, just like the telecommunications sector, will be (completely) liberalized. Concerning the information production framework, the commercial information production would be liberalized as well. However, the public service television's content remains a national competence within the information production framework, because this is part of a nation's cultural policy domain.

14.5   Summary and conclusions

A pessimistic and an optimistic future scenario for society on how interactive DTV services—which are provided via a CA system—are established on a mid-term time scale were described. It is feasible to say that, depending on the society concerned, some DTV aspects' development will have elements of both scenarios in practice. However, an optimistic scenario as a whole is the objective.

Furthermore, the migration paths for the introduction of DTV and CA services were discussed. Concerning CA, it has to be stated that each of the three models—Multicrypt, Simulcrypt, and Transcontrol—has its own advantages and disadvantages. The market will decide which model to use, while the governments will need to ensure an open market structure. With regard to DTV in general, a responsible introduction scenario also requires a feasible time scale, including a simulcasting period. This allows the various market parties to synchronize the introduction of their products and services for DTV. Meanwhile, governments have sufficient time to allocate the required frequencies for satellite and terrestrial DTV services.

Finally, the future will bring no extrapolation of past developments in television, but rather a paradigm shift as a result of a convergence process. Such convergence not only refers to newly arising multimedia services and converging traditional sectors but also to (near) future dramatic changes in the existing policy and regulatory frameworks.

References

[1]   de Bruin., R., *Technologie Beleidsonderzoek naar Interactieve Digitale Video-diensten met Conditional Access*, Technische Universiteit Eindhoven, October, 1995.

[2]   Verdict. 16 August 1996, rolenumber KG 96/2275G (Summary proceedings by the Consumers' Organization, H.W.A. Kiel, J.H. Hoeflaken and E.G. Lantink versus RTL/Veronica De Holland Media Group S.A., CLT S.A., RTL4 S.A., Veronica RTV Beheer B.V., Scandinavian Broadcasting System SBS6 B.V. and Multichoice Nederland B.V.).

[3]   de Bruin, R., "Making Interactive TV Pay," *Telecommunications International*, pp. 105–108.

Glossary

AC-3   audio compression; Dolby standard for multichannel audio source coding

ACATS   Advisory Committee on Advanced Television Service

A/D-converter   analog-to-digital converter

AM   amplitude modulation

API   application protocol interface

ARPA   Advanced Research Projects Agency

ASCII   American Standard Code for Information Interchange

ATA   awareness technology assessment

ATM   asynchronous transfer mode

ATSC   Advanced Television Systems Committee

ATTC   Advanced Television Test Center

ATV   advanced television

AWGN   additive white Gaussian noise

B   bandwidth

BAT   Bouquet Association Table

BB   baseband

BBC   British Broadcasting Corporation

BER   bit error ratio

BNA   broadcast network adapter

BNI   broadband network interface

BOC   Bell operating company

BSS   broadcasting satellite services

BTA   Broadcasting Technology Association

CA   conditional access

CAMS   conditional access management system

CAT   conditional access table

CATV   cable antenna television

CCIR   Comité Consultatif International des Radiocommunications (an ITU organization)

CCIR IWP   CCIR Interim Working Party

CCITT   Comité Consultatif International Télégraphe et Téléphone (an ITU organization)

CENELEC   Comité Européen de Normalisation Electrotechnique

CI   common interface

CLUT   color look-up table

CoJ   court of justice

COM   Communication of the European Commission

CPS   cable programming service

CSA   common scrambling algorithm

CTA   constructive technology assessment

CW   control word

C-MAC   combined multiplexed analog components

D/A-converter   digital-to-analog converter

DAB   digital audio broadcasting

DARPA   Defense Advanced Research Projects Agency

DAVIC   Digital Audio-Video Council

DBS   direct broadcast satellite

DCT   discrete cosine transformation

D-MAC   duo-binary multiplexed analog components

D2-MAC   duo-binary multiplexed analog components; "D2" indicates that the bit rate is half of the D-MAC standard

DEMUX   demultiplexer

DIAMOND   digital television project within SPECTRE

DIGSMATV   digital television project within RACE

DSB   digital satellite broadcasting

DSM-CC   digital storage media command and control

DTH   direct-to-home

DTS   decoding time stamp

dTTb   digital television project within RACE

DTV   digital television

DVB   digital video broadcasting

DVB-C   DVB cable specification

DVB-CI   DVB common interface specification

DVB-CS   DVB satellite master antenna television specification

DVB-MC   DVB multipoint video distribution system specification

DVB-MS   DVB microwave multipoint distribution system specification

DVB-NIP   DVB network independent protocols specification

DVB-RCC   DVB return channel through CATV networks specification

DVB-RCT   DVB return channel through PSTN/ISDN specification

DVB-S   DVB satellite specification

DVB-SI   DVB service information specification

DVB-SIM   DVB simulcrypt specification

DVB-SUB   DVB subtitling specification

DVB-T   DVB terrestrial specification

DVB-TXT   DVB teletext specification

EBU   European Broadcasting Union

E-cash   electronic cash; an electronic equivalent of cash

ECM   entitlement control message

ECU   European currency unit

EDI   electronic data interchange

EDTV   extended definition television

EEC   European Economic Community

EIT   event information table

ELG   European Launching Group

E-mail   electronic mail

EMI   Electric Musical Industries

EMM   entitlement management message

EN   European norm

EPG   electronic program guide

ETR   European technical requirements

ETS   European technical standard

ETSI   European Telecommunications Standards Institute

FCC   Federal Communication Commission

FDMA   frequency division multiple access

FEC   forward error correction

FIFO   first-in-first-out

FSS   fixed satellite services

GA   Grand Alliance

GB   guard band

GSM   global system for mobile communications

HD-DIVINE   Scandinavian project on terrestrial HDTV

HD-MAC   high-definition multiplexed analog component

HDTV   high-definition television

HVC   High-Vision Promotion Center

IAB   Internet Architecture Board

IB   in-band

IC   integrated circuit

ICT   information and communication technology

ID   identification

IDTV   interactive digital television

IEC   International Electrotechnical Commission

IESG   Internet Engineering Steering Group

IETF   Internet Engineering Task Force

IF   intermediate frequency band (0.95 GHz–2.05 GHz)

INA   interactive network adapter

INI   interactive network interface

IRD   integrated receiver decoder

ISDN   integrated services digital network

ISO   International Standards Organization

ISOC   Internet Society

IT   information technology

ITJ   International Telecom Japan

ITU   International Telecommunications Union

ITU-D   ITU development sector

ITU-R   ITU radiocommunication sector

ITU-T   ITU telecommunications standardization sector

IWP   Interim Working Party

IXC   inter exchange carriers

JTC   Japan Telecom

JTC1   (ISO) Joint Technical Committee 1

KDD   Kokusai Denshin Denwa

LEC   local exchange company

LEO   low earth orbit

LFSR   linear feedback shift register

LRIC   long-run incremental cost

MAC   media access control

MAC   multiplexed analog component

MHP   multimedia home platform

MITI   Ministry of International Trade and Industry

MMDS   microwave multipoint distribution service

MN-HDTV   MUSE decoder NTSC-HDTV

MoU   memorandum of understanding

MP@ML   main profile at main level

MPEG   Motion Pictures Experts Group

MPEG-1 to 4   MPEG standards on video coding, service information, and multiplexing

MPEG layer I to IV   MPEG standards on audio coding

MPT   Japanese Ministry of Post and Telecommunications

MSB   most significant bit

MUSE   multiple sub-Nyquist sampling encoding

MUX   multiplexer

MVDS   multipoint video distribution system

NHK   Nippon Hoso Kyokai

NAB   National Association of Broadcasters

NIT   network information table

NIU   network interface unit

NSF   National Science Foundation

NTSC   National Television System Committee

NTT   Nippon Telegraph and Telephone

NTV   Nippon Television Network

OFDM   orthogonal frequency division multiplex

OH   overhead

OJ   official journal

ONP   open network provision

OOB   out-of-band

OSI   open systems interconnection

PAL   phase alternation line

PAT   program association table

PC   personal computer

PCMCIA   Personal Computer Memory Card International Association

PCR   program clock reference

PCS   personal communication system

PDH   plesiochronous digital hierarchy

PES   packetized elementary stream

PID   packet identification

PLL   phase-locked loop

PMT   program map table

PRBS   pseudo random binary sequence

prETS   pre-European technical standard

PPV   pay-per-view

PRBS   pseudo-random binary sequence

PS   program stream

PSI   program-specific information

PSK   phase shift keying

PST   public service television

PSTN   public switched telephone network

PTS   presentation time stamp
partial transport streams

PTT   Post Telegraph and Telephone

Q   quantizer

QAM   quadrature amplitude modulation

QEF   quasi-error-free

QPSK   quadrature phase shift keying

RACE Research and Development in Advanced Communications Technologies in Europe

RB reference burst

RCA Radio Corporation of America

RF radio frequency

RS Reed-Solomon

RSA Rivest, Shamir and Adleman's public crypto system

RST running status table

SAS subscriber authorization system

S-band super band (0.23 GHz–0.47 GHz)

SDH synchronous digital hierarchy

SDT service description table

SECAM Système Électronique Couleur Avec Mémoire

SFN single frequency network

SI service information

SMATV satellite master antenna television

SMS subscriber management system

SNR signal-to-noise ratio

SPECTRE experimental European research program

ST stuffing table

STA strategic technology assessment

STC system time clock

STERNE digital television project within SPECTRE

STU set-top unit

SYNC synchronization

TA technology assessment

TB   traffic burst

TBL   telecommunications business law

TCP/IP   transmission control protocol/Internet protocol

TDMA   time division multiple access

TDT   time and date table

TOT   time offset table

TPS   transmission parameter signaling

TS   transport stream

TT   teletext (*see also* TXT)

TTP   trusted third party

TV   television

TVWF   television without frontiers

TXT   teletext (*see also* TT)

UHF   ultra high frequency band (0.3 MHz–3 GHz)

UN   United Nations

VAN   value-added network

VBI   vertical blanking interval

VSB   vestigial sideband

WRC   World Radio Conference

WTO   World Trade Organization

About the authors

Ronald de Bruin was born in the Netherlands. He studied electronic engineering at the Rotterdam Polytechnic and technology policy sciences at the Eindhoven University of Technology. He has served as a telecommunications policy advisor in the field of information security and data and privacy protection at the Dutch Telecommunications and Post Department. One of his accomplishments there was the establishment of a policy framework on trusted third parties (TTPs). He is currently active as a manager at KPMG TTP Services. In addition, he has published several articles on conditional access and TTPs, and at present, he is preparing his dissertation on the application of TTP services in conditional access systems.

Jan Smits holds a Masters (LL.M) from Tilburg University, and a PHD in law from Utrecht University. In 1992, he was appointed as chair of law and technology in the Technology Management Department of the Eindhoven University of Technology. In addition, he is involved in telecommunications consultancy both as an independent consultant and as an associate to KPMG. From 1993 to its abolishment in 1995, he was a member of the Media Advisory Council to the Dutch government. He has also been an advisor to the ITU and UNDP on telecommunications and development issues. He has published writings on artificial intelligence and law issues, as well as numerous books and articles on international telecommunications policy and regulation. In 1991, Nijhoff Publishers released a commercial publication of his Ph.D. research *Legal Aspects of Implementing International Telecommunications Links*. In addition, Smits co-authored with Rudi Bekkers the 1999 Artech House book *Mobile Telecommunications: Standards, Regulation, and Applications*.

Index

A

Actors
  activities, changes in, 39–44
  defined, 37
  gatekeepers, 39
  layer modeling of, 37–39
  See also Layer model
Additive white gaussian noise
  (AWGN), 168
Administrative guidance, 73–74
Advanced television (ATV), 56
Advanced Television Systems
  Committee (ATSC), 10, 65
Advanced Television Test Center
  (ATTC), 50
Advisory Committee on Advanced
  Television Service
  (ACATS), 49
Affordability, 116–17, 122
  European DVB, 263–64
  optimistic scenario, 275
  pessimistic scenario, 273
Analog-to-digital (A/D)
  converter, 141
Analytical model, 115–24

DVB and, 258–62
introduction, 115–16
technological development
  aspects of, 116–21
Application protocol interface
  (API), 134
ARPANET, 243
Asian Pacific Telecommunication
  Community (APT), 84
Asynchronous transfer mode (ATM)
  networks, 134
Audio coding, 142–46
  bit stream, 144, 145, 146
  decoding, 145–46
  DVB guidelines for, 146
  encoding, 142–45
  MPEG layer I encoding
    system, 143
  MPEG layer II encoding
    system, 144
  MPEG layer I/layer II decoding
    systems, 146
  sampling rates, 146
  See also Coding; MPEG-2 standard
Availability, 116

The Artech House Digital Audio and Video Library

Communication and Computing for Distributed Multimedia Systems,
Guojun Lu

Computer-Mediated Communications: Multimedia Applications,
Rob Walters

Digital Video Broadcasting: Technology, Standards, and Regulations,
Ronald de Bruin and Jan Smits

Digital Video Communications, Martyn J. Riley and
Iain E. G. Richardson

*Distributed Multimedia through Broadband Communications
Services,* Daniel Minoli and Robert Keinath

Fax: Facsimile Technology and Applications Handbook,
Kenneth McConnell, Dennis Bodson, and Richard Schaphorst

Networks and Imaging Systems in a Windowed Environment,
Marc R. D'Alleyrand, editor

Packet Video: Modeling and Signal Processing, Naohisa Ohta

Principles of Digital Audio and Video, Arch C. Luther

Television Technology: Fundamentals and Future Prospects,
A. Michael Noll

Video Camera Technology, Arch C. Luther

Videoconferencing and Videotelephony: Technology and Standards,
Richard Schaphorst

For further information on these and other Artech House titles, including
previously considered out-of-print books now available through our
In-Print-Forever™ (IPF™) program, contact:

Artech House
685 Canton Street
Norwood, MA 02062
781-769-9750
Fax: 781-769-6334
Telex: 951-659
e-mail: artech@artech-house.com

Artech House
Portland House - Stag Place
London SW1E 5XA England
+44 (0) 171-973-8077
Fax: +44 (0) 171-630-0166
Telex: 951-659
e-mail: artech-uk@artech-house.com

Find us on the World Wide Web at:
www.artech-house.com

DATE DUE

DEMCO 38-297